如何写出高水平英文科技论文——策略与步骤

（原著第二版）

Writing Scientific Research Articles

STRATEGY AND STEPS

（澳）玛格丽特·卡吉尔 （澳）帕特里克·奥康纳 著
(Margaret Cargill) (Patrick O'Connor)

龚伟峰 译

化学工业出版社

·北京·

本书作者从策略、论文叙述、语言三个方面，分别介绍了英语论文的写作技巧，包括科技论文的结构、审稿人评判标准、论文各部分写作技巧、如何回复编辑与审稿人等，内容详细、实用。

本书针对的主要读者群是以英语作为第二语言的科技工作者，对于科研人员、在校大学生、研究生等都有很好的参考价值。

Writing Scientific Research Articles：Strategy and Steps，Second Edition/by Margaret Cargill and Patrick O'Connor

ISBN 978-1-118-57069-2（cloth)-ISBN 978-1-118-570570-8（pbk.）

北京市版权局著作权合同登记号：01-2018-4193

图书在版编目（CIP）数据

如何写出高水平英文科技论文：策略与步骤/(澳)玛格丽特·卡吉尔(Margaret Cargill)，(澳)帕特里克·奥康纳(Patrick O'Connor) 著；龚伟峰译. —北京：化学工业出版社，2018.6（2022.11重印）

书名原文：Writing Scientific Research Articles：Strategy and Steps，2nd Edition

ISBN 978-7-122-31890-9

Ⅰ.①如… Ⅱ.①玛…②帕…③龚… Ⅲ.①科学技术-英语-论文-写作 Ⅳ.①G301

中国版本图书馆 CIP 数据核字（2018）第 065457 号

责任编辑：韩霄翠　梁玉兰　仇志刚　　文字编辑：高　宁
责任校对：宋　玮　　装帧设计：尹琳琳

出版发行：化学工业出版社（北京市东城区青年湖南街 13 号　邮政编码 100011）
印　　装：天津盛通数码科技有限公司
787mm×1092mm　1/16　印张 13　字数 244 千字　2022 年 11 月北京第 2 版第 4 次印刷

购书咨询：010-64518888　　售后服务：010-64518899
网　　址：http://www.cip.com.cn
凡购买本书，如有缺损质量问题，本社销售中心负责调换。

定　　价：48.00 元

前言

本书第一版的读者遍布世界各地。21世纪，科研生涯与论文发表密切相关，无论是对于科研人员中的新手作者，还是扮演指导角色、帮助这些人实现上述任务的导师或语言教师，我们的这本书都起到了良好的辅助作用，受到读者欢迎。本书初版的许多使用者为我们提供了建议，在教学中我们也积累了新的经验，这些内容都会在第二版中有所体现。除此之外，我们还在第二版中探讨了一些新的话题。

首先，在第2章原有基础上，我们补充了科技论文的一种常见结构（AIBC），它适用于物理、计算机科学以及其他的一些工程领域，书后还相应增加了一篇该类型的论文范例。这样一来，第2章就覆盖了科技论文中所有常用的宏观结构类型，能够满足不同学科研究者的写作需要。

其次，我们新增了一章介绍综述的撰写。事实上，发表综述与发表论文的策略并无二致。你只需要转换思路，分清综述中“数据”和“关键信息”的内涵，前者代表综述作者对文献资料的评估，后者则代表综述要提出的新动向、新观点。然后借鉴论文的写作原则准备稿件即可。

最后，为了满足读者的需求，我们再一次采用类似的策略分析了如何准备科研项目申请书。不同的基金组织、资助单位评判项目的具体标准不同，但申请人需要了解相关标准，并在申报项目时有效地回应它们，这个过程是不变的。你要体会的是，如何把本书在其他章节提出的策略和建议灵活运用到不同的情境中，小到一份出差、参会申请，大到一项国家、国际基金项目，这些原则都可以帮助你从容应对。

感谢 Holly Slater、Andrew Smith、John Harris、Peter Langridge、Matt Gilliham 和 Michelle Picard 等同事在第二版成书过程中给予我们的帮助，感谢 Wiley-Blackwell 出版集团的编辑 Ward Cooper、Carys Williams 以及 Kelvin Matthews 的辛勤付出。衷心感谢我们第一版图书的读者，你们的想法和疑问是我们完成这本书的动力。按照惯例，欢迎本书的使用者提出批评意见，以便我们今后修订时参考。

Margaret Cargill
Patrick O’Connor
2012年9月

第一版前言

刚刚起步的科研工作者在把研究结果转化成文字，进而发表的过程中总会感到力不从心；许多人同样也会好奇：在国际学术期刊中发表英文论文，如何才能让成功率更高，有哪些技巧可以参考。这些都是本书要回答的问题。几乎所有的科研人员都希望能在高水平期刊上发表文章，但是具体到写作和发表又面临着压力和挑战。我们认为，想要成为更有经验的论文作者，最实用的方法是从三个方面取得突破：

- 掌握写作策略：了解编辑和审稿人希望发表什么样的文章，以及背后的原因
- 讲好研究故事：特定学科内那些引人入胜的文章，它们最终得以发表，这些文章的共同特征是什么
- 精炼论文语言：如何写出更清晰、更有效的英文，与读者进行沟通

这其实不难理解，想要把成功的可能最大化，就需要在科学层面和语言层面共同努力。本书的作者团队也是这样搭配的：一位是科研人员，一位是语言教师，我们把各自的视角和经验结合起来，为攻克英文论文写作搭建了一套完善的跨学科体系。

无论英语是否是你的母语，你都可以借鉴本书的内容。当今世界，大多数科技论文都是用英语写成的，但论文的作者却具有不同的语言、文化背景，科研项目也更加国际化，涉及大量的跨文化交流。在这一背景下，论文作者发表文章将面临诸多挑战；除此之外，期刊的编辑、审稿人，教师以及其他为论文发表提供支持的工作者面临的任务也更加严峻。如果你需要参与科技论文的写作，希望本书能够带给你一些收获。

本书既适用于自学，也可作为教学材料，供教师在课堂上使用。读者可以一边学习本书，一边有序准备自己的手稿；另一种选择是在准备阶段通读全书，然后根据写作进度复习相关的章节，逐步实现发表。我们还在网站上（http://www.writeresearch.com.au）提供了丰富的学习资源。

本书的完成得益于我们与阿德雷德大学（University of Adelaide）多年的合作，还得益于2001年以来我们与中国科学院的合作。我们要感谢许多人，首先是 *Writing Up Research: Experimental Research Report Writing for Students of English* 一书的两位作者 Robert Weissberg 和 Suzanne Buker，这部1990年出版的作品开创性地使用了语类分析理论来剖析科技论文的语言，为论文写作的教学奠定了坚实的基础。还要感谢 John Swales

和他的团队，他们多年以来形成的研究成果、理论模型和教学材料有助于我们深入了解不同学科论文的风格。基于上述研究成果，我们在阿德雷德大学的团队才有机会开展一系列的研究和实践，我们要特别感谢 Kate Cadman、Ursula McGowan 和 Karen Adams 等多位科学家的参与和付出。在中国开展教学活动期间，许多资深的作者都与我们分享了他们的视角和经验，在这里尤其要感谢 Andrew Smith、Brent Kaiser、Scott Field、Bill Bellotti、Anne McNeill 以及 Murray Unkovich。同时，我们对 Yongguan Zhu、Jinghua Cao 以及其他支持我们培训项目的单位和个人表示感谢，经过多次的教学实践，我们才得以不断改进教学方法。感谢我们研讨会的所有参与者，我们的学员来自澳大利亚、越南、西班牙和中国，学员们贡献的建议和热情能够使本书更加贴近读者的需求。

最后，衷心感谢 Sally Richards、Karen Adams、Marian May 以及 Wiley-Blackwell 出版集团的编辑 Delia Sandford 和 Ward Cooper 在出版过程中给予我们的帮助。书中难免存在疏漏之处，请读者不吝赐教。

Margaret Cargill

Patrick O’Connor

2008 年 9 月

目录

第一部分 01

明确论文框架，方能事半功倍

1. 本书的结构与思路

1.1 写作亦是沟通

欢迎你开启论文写作与发表之旅，将你的研究结果呈现在国际学术期刊中。即便英语不是你的母语，亦或你暂时还只是新手作者，你都可以通过阅读本书，迅速获取科技论文的写作技巧与投稿策略。

在各种参考书或网站上，你会找到许多相关信息——建议你如何准备手稿、写好期刊文章或科研论文（在本书的第 2 章，你将看到科技论文的各种结构类型；第 12 章讲解了综述类文章；第 18 章涉及科研项目申请书）。无论你把即将投出的稿件称作什么，希望你能意识到上述概念都指向同一种文本类型，即科技论文这一语类。本书汇集了众多研究者在针对科技论文进行语类分析（也称体裁分析）时得出的结论，这些结论可以帮我们更清晰地了解科技论文的本质特征。

为找到高效写作的捷径，研究者们提出了许多核心概念，目标读者就是其中之一。无论你所草拟的文件是何种类型，在任何时候都应优先考虑目标读者。目标读者也是“沟通魔方”的四要素之一，其他要素分别是写作目的（你希望自己的文章达到何种目标）、写作格式（既定格式会对你的写作产生哪些限制）以及评价指标（如何界定文章是成功的）。本书会借助“沟通魔方”中的全部要素对论文进行分析。首先我们就来分析科技论文的目标读者。

谁是你的目标读者?

你率先想到的应该是你的同行：同行们与你从事的研究领域密切相关，希望了解你的研究进展，是你论文的主要读者。但请不要忘了，在这些人有机会读到你的文章以前，你必须先达到期刊编辑和审稿人的要求（详见第 3 章、第 13 章、第 14 章），编辑、审稿人作为期刊的“过滤器”，只会发表那些符合条件的文章。因此，从准备写作的阶段开始，你就有必要了解编辑和审稿人的要求，并把这些要求铭记于心。本书归纳出了审稿人评判稿件的标准（详见第 3 章、第 14 章），并以这些标准为框架，对科技论文进行剖析，向你展示如何才能满足目标读者们的期待。这涉及如下两个方面：

- 论文各部分的内容和呈现方式；
- 用来呈现论文内容所使用的专业英语。

上述两个方面密切相关，为帮助你同时满足要求，本书采取跨学科的视角：在内容方面，参考有经验的作者和审稿人的建议；在语言方面，借鉴语类分析的研究成果。在分析论文各部分内容的同时，我们将其核心语言特征融入相同章节一并介绍。针对母语非英语的作者，本书在第 17 章专门讲解了提升专业英语技能的方法，你可以在完成本章的学习之后，自行决定学习该部分内容的时机。

1.2 顺利发表为何不易

在准备开始写作之前，不妨先思考如下问题：为什么要发表？发表文章为什么很难？如何融入国际科学界？如何选择目标期刊？如何充分利用发表的机会？我们的解答如下。

为什么要发表?

上文已经提到的一个原因是：发表论文可以跟同行分享观点、汇报研究结果。此外，发表论文还可以：

- 留下记录，供其他研究者跟进；
- 得到同行认可；
- 吸引当前领域内其他研究者的兴趣。

通过在国际期刊中发表论文，尤其可以：

- 得到专家对论文观点和研究结果的反馈意见；
- 将研究方法和结果交由编辑和审稿人检验，使文章达到既定标准。

上述原因也刚好体现了审稿程序的重要性。然而，论文得以成功发表并非易事，所有的科研工作者都面临着某些困难，母语非英语的作者面临的困难则更多。

顺利发表为何不易?

除了语言障碍，你还应该意识到，无论使用何种语言，写作本身也是一门技能。因此，即使英语是你的母语，本书提供的许多写作建议仍值得关注。

成功发表论文也有技巧可寻。不是每位作者都有发表机会，原因如下。

- 某些研究的创新性或学术价值不足。
- 实验总会有失败，而正面的研究结果更容易获得发表机会。
- 出版行业是买方市场：期刊能针对稿件提出各种要求，某些要求作者可能无法满足。

在后面的章节中，本书将对此提供合理的建议。

另外，研究者通过发表论文向同行介绍研究工作的同时，也面临着被批判的风险。推进知识的进步、观点或概念的演化总会遭遇反对的声音，新的观点、结果在得到认可前，都要经过严格的论证。面对空白的页面，想到挑剔的读者，研究者定能感受到论文写作与发表的挑战性。本书针对论文的每个部分都提供了思考和写作的框架，帮助你理清思路，分解写作任务，应对发表过程。

如何融入国际科学界?

向国际期刊投稿就是在融入学术界，你其实是在努力与国际同行进行交流。为了获得交流资格，你需要了解其他人已经发表过哪些观点，也就是说，你需要站在当前研究领域的最前沿，对国际范围内领军人物的研究成果了如指掌。这就要求你：

- 阅读当前研究领域内的相关期刊；
- 登录期刊网站，使用邮件提醒功能，定期查收新刊目录；
- 熟练掌握网络和图书馆电子数据库的检索功能。

否则，你将无法准确界定你的研究工作，使之与领域内的最新进展相匹配。事实上，你在着手写作前，甚至在确立研究计划前，就应该了解研究领域的现状，这样才能确保你即将开展的研究可以与国际同行“对话”。

积极参加国际学术会议也是帮助你融入的重要途径。因此，与同行交流既需要写作技能，也需要口语技能。本书侧重写作技能，但我们在第16章列出了一些提高口语交流能力的建议，供你参考。上述准备工作其实是在帮助你进行学科知识的建构，然后，你将具备筛选目标期刊的能力，确定自己会将手稿投向何处。

1.3 目标期刊的选定

目标期刊会决定你顺利发表的速度和难度。在研究工作开始之时，你就应思考适合的目标期刊；在着手引言、讨论部分的写作时，你应确定自己的选择。

目标期刊应刚好面向你的目标读者群体，为你在学术领域带来声誉，并能在刊发速度和难度上达到平衡。为提高录用概率，建议你定好方案，一旦目标期刊退稿，可转投其他备选期刊。在选定目标期刊时，请考虑下列因素。

- 该期刊是否会发表你所从事的那类研究工作。查阅几期该刊的内容，如果期刊有网站，也可搜索相关信息。在你论文的引言部分引用该刊的文献会非常有帮助，这说明你加入到了期刊正在进行的“对话”中。观察你在引言部分使用的主要参考文献，了解它们在引言中引用了哪些期刊。用这种方法不断对文献进行追溯，可以帮助你迅速从整体上把握当前领域的期刊情况，而你在引言、讨论部分引用最多的期刊，则更有可能发表你所从事的那类研究工作。

- 该期刊的出版宗旨和范围是否与你的研究内容、研究方向在学科中所处的层次相匹配。查阅期刊网站上关于aims and scope的说明，阅读发表在该刊上的文章，找到最适合自己手稿的那种刊物。这样做还可以确保文章在发表后，可以直达你设想的受众。
- 在该期刊上发表文章能否满足你的个人需求。首先要确认该刊是否会对来稿进行评审。因为经过同行评议的文章无疑具有更高的信度。其次，要确认该期刊的影响因子。许多国家或研究团队在评估研究人员的成果时，也会对期刊的影响因子有要求（关于期刊质量的各种评价指标，详见本书附录）。
- 该期刊的发表周期是否合理。许多刊物都会在论文标题的下方注明收稿日期和发表日期，你可以据此进行分析。一些期刊的网站上也能找到相关内容。某些期刊会在发表纸质版之前，将文章先一步刊登在网站上，通常情况下，此类刊物发表周期更短。期刊本身也希望缩短发表周期，这样做能更好地吸引从事创新型研究的作者。如果有研究者与你从事着类似的工作，快速发表可确保你占得先机，提高引用率，这对于晋升或经费申请都有帮助。
- 该期刊是否收取版面费。提前确认该刊在发表过程中或印制彩版时是否会收取费用；你还可以询问编辑部是否愿意为某些地区的作者减免发表费用。你或许还希望自己的文章可以覆盖更多无法利用图书馆资源以及未付费订阅该刊的读者。目前，许多期刊都支持开放存取（open access，OA），即发表付费，阅读、下载免费，如果你所在的研究机构或你个人有类似需求，可以了解选定的期刊是否提供相关服务。
- 该期刊编辑人员的工作效率以及是否愿意提供帮助。一些期刊会在官网上为母语非英语的作者提供有针对性的建议。你还可以请教有过该期刊投稿经历的同事。在实验室或团队内部分享相关信息十分有益，这有助于你所在的研究机构或团队在国际期刊中发表更多成果。

如何充分利用发表的机会?

将研究成果迅速发表对你而言非常有利。此外，目标期刊应拥有广泛的读者群（是引文索引中高水平、高影响力的期刊，相关说明参见附录）。但如果你想追求国际影响力而把目标定得过高，则可能遭到拒稿，拖慢发表进度。因此必须结合自身情况，在刊发速度和难度上取得平衡，制订最优策略。最后，建议你发表在同行经常阅读的刊物上。请完成任务1.1，回答问题并填写表格，梳理你对上述内容的理解。与你论文的合著者讨论表中信息，制订发表策略。

在完成以上内容后，你应已初步选定了目标期刊，接下来为你介绍本书的目标。

任务 1.1 分析备选的目标期刊

为充分利用发表的机会，建议你制定发表策略，其中重要的一环就是选定目标期刊。首先，选出三到四种备选期刊，然后回答下列问题，将答案记录在表 1.1 中。

1. 在过去三年内，期刊是否发表过与你的研究工作类似，具有相似创新水平的文章？如果答案为"否"，那么在向该期刊投稿前应慎重考虑。

2. 期刊近期发文的范围和内容是否与你的手稿在主题、方法、结果方面相匹配？在表格内填入该刊发表文章的主要类型，如"plant physiology：non-molecular studies"。

3. 对你的研究领域和你个人而言，有关期刊质量、影响力的哪个指标最重要？针对每本期刊，在表格中记录具体评价指标（如"影响因子或被引半衰期"）或相关数值。

4. 期刊的发表速度如何？相关信息可在网站或每篇文章内找到。写下具体时间，或针对发表周期对期刊进行评分（如"录用后三个月内发表＝快；周期超过一年＝慢"）。

5. 期刊收取版面费吗？提供开放存取服务（OA）吗？是否额外收费，以及你能否承担相关费用？

衡量你在表 1.1 中打出的分数，对期刊的总体评价进行排名。

表 1.1 结合关键指标为备选期刊打分

序号	期刊名称	近期发表过创新水平相当的类似研究	是否匹配你近期研究工作的范围、内容	质量、影响力	发表周期	版面费或OA费用
1						
2						
3						
4						

1.4 本书的目标

本书旨在帮助你：

- 进一步理解发表在国际期刊中的英文论文的结构和深层逻辑；
- 学会如何将一系列研究结果转化成可以发表的论文；
- 掌握自己研究领域论文的结构和语言特征的分析方法，并据此来修正手稿；
- 熟悉投稿流程，学习不同发表阶段的应对策略；
- 学习并初步掌握论文各个部分常见的语言特征；

- 获得改进手稿的策略与工具，包括要点核查清单、对语言进行再利用的方法、语料库检索软件以及构建本专业的写作小组；
- 按照目标期刊的要求，完成手稿。

1.5 本书的结构

行动是最好的学习方法。即使一开始本书是你使用的随堂教材，你或许也希望能够通过自身努力以及与同事合作，在课堂之外持续取得进步。因此，本书旨在向你展示如何借助期刊论文学习写作，这项技能不止可以帮助你顺利发表自己的文章，也适用于你可能会尝试的其他写作领域。

为熟练掌握这项技能，希望你可以经常与同事一起研讨论文实例，并与其他组员分享讨论结果，这比较适用于教室场景。自学本书时，不妨先独立思考完成习题任务，然后在学完相应的章节后核对答案进行修正。你还可以一边学习本书，一边有序准备自己的手稿，并按照书中的介绍进行润色和校对。

本书涉及的论文实例有如下类型。

- 论文示例（PEAs，provided example articles）：包含三篇完整论文，附在书后（第 19 章～第 21 章），由本书作者提供。在学习前面的章节时，你需要对三篇 PEA 进行分析；随后，我们会要求你选择其中一篇对其进行深入剖析。
- 自选论文（SA，selected article）：由你从自己的研究领域中选出，可以是从目标期刊中选出的文章。请参考任务 1.2 中的建议。
- 论文初稿（OA，own article）：代表你希望基于自己的研究结果写出的论文，可以一边阅读本书，一边修改。如果尚未得出实验结果，可暂时跳过涉及 OA 的练习，等研究结果就绪后，再回过头来学习。

任务 1.2 选出用于分析的论文（SA）

SA 需要来自你的研究领域，最好取自目标期刊，论文的作者也应以英语为母语（可通过查看作者的姓名和工作单位判断其语言背景）。不推荐从“Nature”（UK）或“Science”（USA）中选择 SA，因为这两本期刊中的文章风格与大多数其他期刊不同。应该先从更常规的文章风格学起，如果后期有需要，再进行调整即可（论文的常见结构及变体，详见第 2 章）。

本书其他章节按照以下思路安排。

- 首先呈现论文的常见结构以及论文各个部分的内容要点，这些都是过去 20 年以来，语类分析领域研究成果的结晶。本书只负责“描述”特征，而不是“规定”你该如何去做，毕竟有效的写作方式不止一种。研究人员在分析论文时发现的特征，都可以作为写作策略供你参考，在完成论文的不同部分时，你可以根据具体的写作目标来选用。

- 随后，你可按照书中描述对 PEA 的相关部分进行分析，看能否找到这些特征（书后附有参考答案）。
- 接下来，在 SA 中查找相关特征，并思考这些特征与作者写作目的之间的关系。
- 最后，利用这些特征对 OA 进行润色加工（若尚未得出实验结果，可暂时跳过）。
- 除了分析论文各部分内容的结构特征，本书还将教会你更好地使用专业英语。在论文的每个部分，都涉及相关的分析和训练。如果英语是你的母语，你可以有选择地跳过这些章节的部分或全部内容。
- 结束对论文的分析后，本书会向你展示如何应用同样的原则完成综述的写作。
- 然后聚焦投稿环节，以及如何回应编辑提出的修改建议。
- 第 15 章汇总了论文写作流程，并提供编辑和校对策略。
- 第 16 章侧重技能提升策略，即如何通过自学或写作小组不断提升使用英语进行写作和发表的技能。
- 第 17 章适用于母语非英语的写作者，分析了学术写作过程中的痛点和常见错误类型。你可以随时学习相关内容。
- 考虑到科研经费对当下研究人员的重要性，第 18 章向你展示如何运用本书的策略完成科研项目申请书。
- 本书正文的最后一部分（第 19 章～第 21 章）包含三篇 PEA，你还可以在本书的网站（www. writeresearch. com. au）上查看更多论文示例。
- 书后附有各章练习的参考答案、期刊质量评价指标的介绍和本书的参考文献。

2. 科技论文的结构

本章介绍科技论文的整体结构。1665年，英国刊发了第一期“Philosophical Transactions”，可以说，科技论文的结构就由此形成，相关的写作惯例不断发展，并沿用至今。尽管不同研究领域的论文都有着相似的核心结构，但值得注意的是它们都形成了各自不同的结构变体。因此在确定手稿的结构前，务必要查阅目标期刊的具体要求。

请先完成任务2.1，再进入下一步的学习。

任务 2.1 论文各部分使用的标题和小标题

翻阅书后的三篇PEA（第19章～第21章），快速阅读文中的标题、小标题：

- 每篇论文都是如何组织的？
- 主要用到的标题、小标题都有哪些？请简要记录。

请核对书后的参考答案。

请分析你的SA（自选论文）中标题使用的情况，不妨与同事（同学）的SA进行对比，找出其中的异同。

2.1 常见结构AIMRaD及其变体

请注意，本书的侧重点是基于实验研究的科技论文（第12章还会涉及综述类文章）；在其他的研究范式中，如人文社科领域，会采用不同的论文结构。即便同样属于实验研究，不同的学科、不同的期刊在呈现不同类型的研究活动时，也会偏好不同的论文结构。首先请看最常见的“沙漏型”AIMRaD（Abstract，Introduction，Materials and methods，Results，and Discussion）论文结构（图2.1），针对论文的各个部分，我们配上了简要的文字说明。请在看图时，关注每个部分的形状和宽度变化，不要在意形状的高度。

将一篇实验研究论文的各个部分组合在一起，可以组成沙漏的形状，这有助于我们更生动、形象地记忆每个部分的核心特征。请在读完图2.1的文字说明后，完成任务2.2。

任务 2.2 “沙漏型”结构是否与你的判断一致？

请讨论：图2.1的论文结构是否也能代表你所在的文化背景或研究机构对于科技论文的理解？若不能代表，请画出符合你理解的结构示意图。

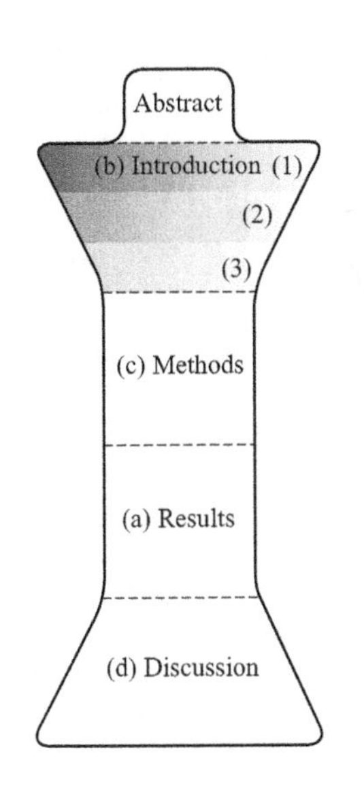

(a) 结果部分的方块决定论文结构图的宽度，论文的全部内容都必须与结果部分中呈现的数据和分析相关联。

(b) (1) 引言的切入点应比较宽泛，旨在吸引目标读者群体以及目标期刊的国际读者。

(3) 引言的结尾应落在对研究结果的关注上，可以介绍研究目的、主要研究工作甚至是研究的主要发现。

(2) 在引言的开头和结尾之间，应将背景介绍、前人研究交织呈现，并在研究空白和当前研究要解决的问题之间建立起逻辑联系。

(c) 方法部分说明研究结果是如何得到的，建立研究的信度。

(d) 讨论部分首先侧重分析当前研究的结果，与结果部分方块的宽度保持一致；但在收尾时，应与引言部分的切入点所涉及的广度一致，即在论文的结尾，说明本研究如何有助于解决论文在开头切入时所提出的更宽泛的议题，从宏观上突出论文的研究工作对当前研究领域的重要性。

图 2.1 “沙漏型” AIMRaD 论文结构及相关说明

当然，并非所有论文都遵循图 2.1 中给出的简单结构。本书将介绍三种 AIMRaD 结构的主要变体（图 2.2～图 2.4），即 AIRDaM、AIM（RaD）C 以及 AIBC。请先读图，然后完成任务 2.3。

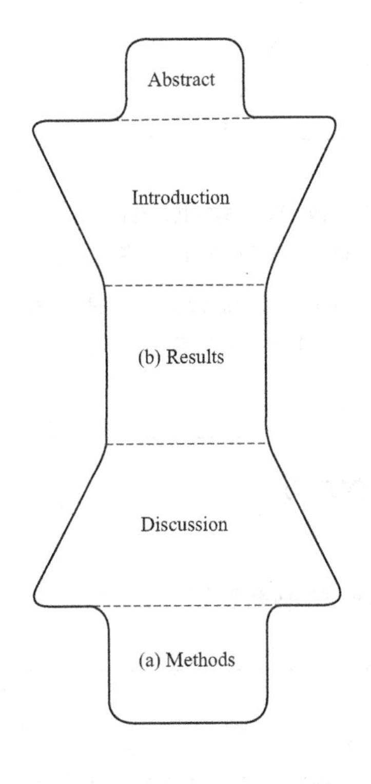

(a) 方法部分（通常改称 Procedure 或 Experimental 部分），置于讨论部分之后，有时使用比正文更小的字号。

(b) 上述变化意味着在结果部分应加入更多实验细节，帮助读者理解实验结果的获得过程。

图 2.2 AIRDaM (Abstract, Introduction, Results, Discussion, and Methods and materials)：常见于化学、分子生物学领域中的论文结构变体

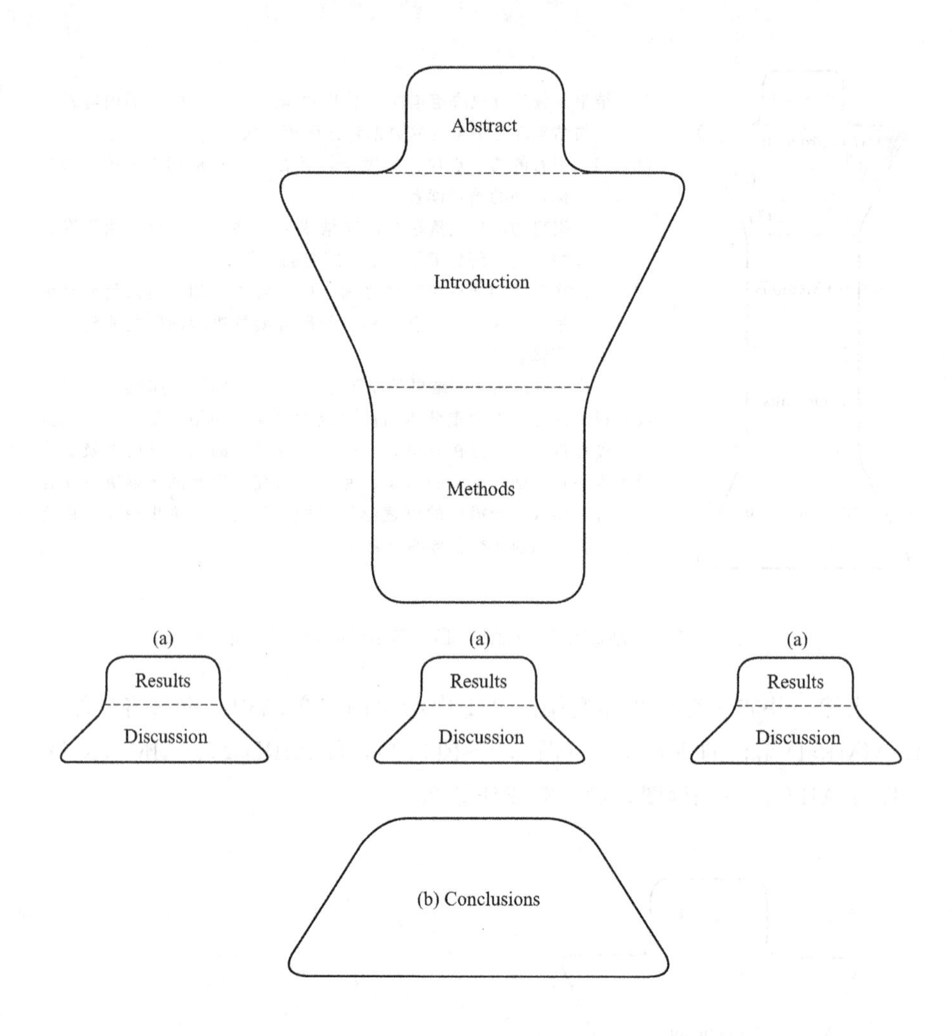

（a）结果和讨论融合为一个部分；在介绍每个结果后，立刻对其进行讨论。
（b）上述变化意味着在论文结尾可加入结论部分，把不同段落中的讨论内容进行整合。

图 2.3 AIM (RaD) C (Abstract, Introduction, Materials and methods, repeated Results and Discussion, Conclusions)：常见于较短篇幅论文的结构变体

任务 2.3 PEA 的结构

结合你在任务 2.1 中记录的答案，请分析：

- 三篇 PEA 分别属于那种论文结构？请核对书后的参考答案。
- 哪种论文结构与你的 SA 最为契合？

科技论文的其他格式

具有高引用率的“Nature”（UK）和“Science”（USA）杂志采用异于上述结构的论文格式，可以看出它们针对的读者群未必是相关领域中的资深研究者，而是更倾向于呈现科学领域的最新重大进展。论文内容经过精心组织，在开头介绍背景信息和开展研究工作的根本原因，面向的读者

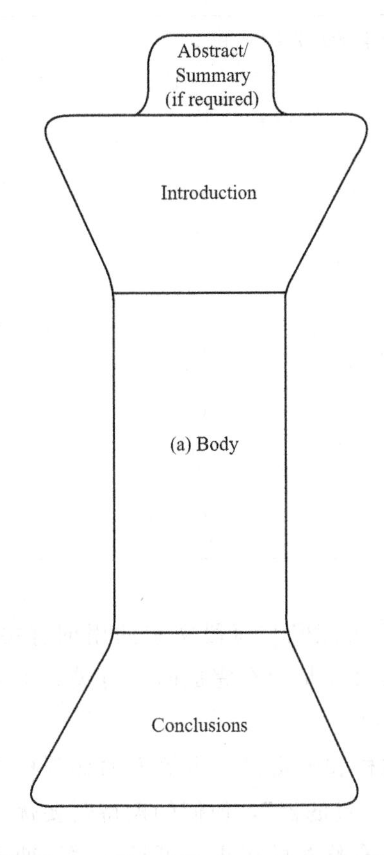

(a) 正文部分包含一系列子部分，按不同主题对标题、小标题进行命名，每个标题都由当前论文内容决定。

图 2.4 AIBC (Abstract, Introduction, Body sections, Conclusions)：常见于工程、遥感、物理、计算机科学领域中的论文结构变体(正文部分可以包含理论、方法、结果、讨论等内容，但根据当前论文特有的研究思路对标题或小标题命名)

群体也更为广泛；随后简要地向读者汇报研究发现并进行简短的讨论。方法部分通常只在正文中概述，详细信息会链接到网站上。此类论文的结构要求也可在期刊官网上获取。在上述期刊中发表论文将面临激烈竞争，对于大多数刚刚起步的研究人员，不适合将其选为目标期刊。因此本书不会过多关注此类论文结构。

除了发表长篇论文，许多期刊还支持发表研究简报（brief notes, research notes 或 notes）和快报（letters）。此类文章不会包含小标题，但如果你仔细分析，依然能发现文中暗含 AIMRaD 以及其他变体的结构。

下面请开始思考，在论文的各个部分应该包含哪些信息？或许你通过阅读自己研究领域的论文，已经可以说出许多答案，不妨完成任务 2.4，测试相关知识。

任务 2.4　预测

请分析下面的文字分别出自论文的哪个部分。在括号中填入相应的字母：I = Introduction，M = Materials and Methods，R = Results，D = Discussion（在 AIBC 型

论文中，上述内容可能会包含在以不同主题命名的小标题之下)。

示例：It is very likely that ... because ... (D)

... yielded a total of ... ( )

The aim of the work described ... ( )

... was used to calculate ... ( )

There have been few long-term studies of ... ( )

The vertical distribution of ... was determined by ... ( )

This may be explained by ... ( )

Analysis was carried out using ... ( )

... was highly correlated with ... ( )

请核对书后的参考答案。

在完成任务 2.4 时，你的判断依据很可能既与每段文字的用词有关，也与它们的语法特征比如时态（一般过去时还是现在完成时）有关。本书会在后面的章节中帮你进一步完善这些知识。

在下一章中，我们希望你能在论文结构和编辑以及审稿人的评判标准之间建立联系，毕竟，这些读者是期刊的“过滤器”，他们的期待需要优先得到满足。本章中涉及的所有论文结构都有悠久的历史，可见，对于期刊而言，这些结构对编辑的工作是有利的，否则这些写作格式也不可能一直保留至今。从这个角度思考，可以带给我们许多有趣的启示。

3. 审稿人评判稿件的标准

正如第 1 章所述，期刊编辑和审稿人是在你投出稿件后的首批读者。近些年来，论文手稿可以通过电子方式提交，因此有时最先读到你文章的人，可能会负责对你上传的文件进行格式和其他方面的检查，但无论如何，仍然是由编辑人员对文章内容进行初步的评估和“过滤”。若确定文章适合送审（详见第 13 章、第 14 章），你的稿件会送交审稿人，审稿人通常有两位。审稿人或许从事着与论文作者相似的研究，或许就是论文某篇参考文献的作者。但评审过程是盲审，也就是不对论文作者公布审稿人的身份（双盲评审是指审稿人也不知道论文的作者是谁，这在科学领域并不常见）。

每种期刊对于审稿人都有专门的要求，这些要求有时可以在期刊官网上查到。建议你查阅目标期刊的网站，留存并参考有关文件。本书整理了一系列相关标准，这些内容都是审稿人在评判稿件时经常需要回答的问题（图 3.1）。除了回答图中“是或否”的问题，审稿人还应指出稿件中存在的问题和改进建议，随后才考虑文章是否适合在该刊出版。由于期刊接收到的投稿越来越多，审稿人还可能会受邀对稿件质量和创新性进行直观地打分——例如，当前论文的水平是否排在过去 12 个月内你审阅过所有稿件的前 20%。审稿人会将所有上述评审意见提交给期刊的编辑。为加深你对评判稿件标准的理解，请完成任务 3.1。

科技期刊提供给审稿人的评估表中包含的常见问题

1. 论文的研究问题新颖吗？
2. 论文的研究内容是否有重要意义？
3. 是否适合在本刊发表？
4. 文章的组织结构是否符合要求？
5. 实验方法和对结果的处理是否符合科学规范？
6. 文章得出的结论是否提供了坚实的数据基础？
7. 文章的篇幅是否合适？
8. 是否包含不必要的插图？
9. 是否包含不必要的图表？
10. 图表说明是否充分？
11. 论文标题和摘要是否明确说明文章内容？
12. 是否引用了最近发表的论文，参考文献的信息是否完整，期刊刊名的缩写是否正确？
13. 文章的整体质量如何：优秀、良好还是很差？

图 3.1　审稿人评判稿件时需要回答的问题

任务 3.1 审稿人在关注论文的哪个部分？

读图 3.1，思考审稿人在回答每个问题时，需要在论文的哪个部分寻找证据，得出答案。在每个问题后面写上代表论文相应部分的字母：A，I，M，R，D 以及 Ref（即参考文献）。例如，你可以在问题 5 后面写上"M，R"。同理，在 AIBC 型论文中，上述内容可能会包含在以不同主题命名的小标题之下。

请核对书后的参考答案。

在深入分析论文各个部分时，希望你能同时关注到这些评判标准。同时，我们会为你介绍如何在论文中运用合理的写作思路和清晰的语言突出审稿人关注的要点。

首先来看论文的标题——你的标题能够明确说明文章的内容吗？

3.1 标题是论文内容的特有标签

好的论文标题应能让读者判断论文所属的研究领域，对论文即将讲述的"故事"，即研究结果有所了解，同时在心中形成与之对应的研究问题。详细内容将在本书第 10 章进行讨论，现在请看下面的例子，然后完成任务 3.2。

论文标题：Bird use of rice field strips of varying width in the Kanto Plain of central Japan

可从标题中获得的信息：
论文的关注点是鸟类与稻田。
研究中的稻田宽度是变化的。
稻田带的宽度与使用它们的鸟类数量、类别有关。
研究地点是日本中部。

在心中可能提出的问题：
为何稻田带的宽度是重要变量？
不同的宽度能否影响使用它们的某种鸟类？
若能，哪种鸟类最常使用多宽的稻田？
研究者如何量化鸟类对稻田的使用？
实验值得在其他地方的稻田重复进行吗？

任务 3.2 论文标题信息提取

分析下列标题，并写下你推断出的相关研究信息和研究结果。作为读者，你希望通过阅读这些论文，得到哪些问题的答案？这些问题也代表着不同读者想要阅读某篇文章的理由。

标题 A：Use of *in situ* ^{15}N-labelling to estimate the total below-ground nitrogen of pasture legumes in intact soil-plant systems

信息：

问题：

标题 B：Short and long-term effects of disturbance and propagule pressure on a biological invasion

信息：

问题：

标题 C：The soybean NRAMP homologue，GmDMT1，is a symbiotic divalent metal transporter capable of ferrous iron transport

信息：

问题：

标题 D：An emergent strategy for volcano hazard assessment：From thermal satellite monitoring to lava flow modeling

信息：

问题：

请核对书后的参考答案。

选择一篇 PEA 完成本书后面的分析任务

任务 3.2 中的标题 B～标题 D 取自书后的三篇 PEA。请你基于上述回答，确定哪篇与自己的研究相关或是你更感兴趣的，将它作为剖析对象，完成本书后面章节列出的文本分析任务。在结束本章的学习之前，请完成任务 3.3。

任务 3.3　解锁你的 SA 标题

请参考任务 3.2 中的要求，对自选论文（SA）的标题进行分析。

标题：

信息

问题：

第二部分 02

论文各部分的写作顺序与方法

4. 结果部分是整篇文章的驱动力

结果部分决定着整篇论文的内容和结构，因此在写作伊始，就应该尽可能清楚地知道论文要讲述的“故事”，即研究结果，包含哪些要点。建议你在准备写作之时，首先就要思考如何清晰呈现研究结果，以及准备将这些结果指向怎样的关键信息（take-home messages）或结论。所谓的关键信息，就是你希望读者在读完论文后，仍然记得的信息，甚至会在第二天喝咖啡闲聊时和同事提起的信息。

为了确保研究结果可以清晰呈现，请首先盘点希望加入论文中的图和表。针对每一个图表，根据数据内容列出一两个要点，说明该图表要传达的主要信息。在每一条信息旁，注明此内容可写进论文的哪个部分：结果、讨论，还是两个部分都要涉及。有的信息甚至需要在引言或方法部分就有所提及，在该条信息旁如实标出即可。随后，确定图表在论文中出现的最佳顺序，以确保“故事”的完整性。根据已经列好的信息点，提炼出论文要传达的关键信息。下一步，请与论文的合著者坐下来一起讨论文章要讲述怎样的研究“故事”，目的是决定：

- 哪些数据要包含在论文中；
- 构成研究“故事”的重要信息点有哪些；
- 应包含哪条或哪几条关键信息。

这也是讨论如何对论文进行署名的好机会（详见第 15 章 15.1 节）。还可以根据 1.3 节中的内容，讨论满足要求的目标期刊。请在做好上述准备工作后，再开始论文各部分的具体写作。

在写作准备阶段，任务 4.1 可以帮你进一步理清论文的关键信息。

任务 4.1　论文起草过程中值得关注的问题

结合你即将写入论文的研究结果，回答下列问题。请务必使用英语回答。

1. 我的研究结果讲的是什么？（上限为两句话，请概述要点，不需要介绍背景）
2. 得出这些研究结果意味着什么？（即通过研究结果可以得出怎样的结论）
3. 谁有必要了解这些研究结果？（即论文针对的读者是谁）
4. 目标读者为什么要知道我的研究结果？（即研究结果对当前领域正在进行的研究工作有怎样的贡献，或者，如果相关的研究人员没有读到我的论文，那会出现什么损失）

如果你已经可以结合自己的研究结果回答上述问题，这代表你知道该如何改进图表的呈现方式，使其最大限度为研究结论提供支持。在第 5 章中，本书将进一步讨论如何设计图表，用数据说话。

5. 结果：用数据说话

在科技论文中展示数据是为论文要讲述的“故事”服务的，它们为支持或推翻假设提供证据，将重要的数据以及元数据记录在案。研究者通过核实、分析和展示数据来分享、构建新的知识，使其获得学术界的认可。为此，我们需要合理地呈现所有必要数据，并突出强调最重要的信息点。展示数据的过程也是决策的过程，某些数据或细节最终不会出现在论文中。如果确定在论文中使用图或表，应将它们进行排序和编号，并按相应顺序把图表融入正文内。

许多期刊都允许作者在附录或在线补充数据中提供附加数据，用来进一步支持或拓展论文内容。因此针对文章涉及的每条数据，都有必要决定其呈现位置——是放在论文中，还是上传到在线数据发挥其参考价值。请注意，审稿人会受邀对所有图表的必要性做出评价，这当然也包括补充材料中的数据。

数据的呈现方式会受到不同学科以及个人偏好的影响，也会随着时间的推移发生改变。你还可以随时找到许多建议，告诉你如何设计出清晰、美观的图表，这些建议有时甚至是相互矛盾的。因此在本章，你不会看到制作图表的固定规则，而是会学习如何优化图表，讲好研究“故事”。最重要的一点是图和表需要具有自明性，可以“独立存在”；也就是说，读者不需要借助正文中的描述就能去理解图表信息，图表本身以及表题、图注说明，或符号表、脚注中应包含全部必要信息。

呈现数据时，务必要参考目标期刊的“投稿说明”（即 Instructions to Contributors，Instructions to Authors，Author Guidelines 等相关文件）。有些期刊不会提出详细要求，但总会对格式编排或风格偏好做出说明。另外，还需要参考该刊近期杂志中的图表风格，通过分析论文中使用的数据类型、图或表的选用、不同图形的选用，以及如何在正文和图表说明中分配对于数据的介绍，你将能最大程度地满足期刊的要求。借助你刚刚的分析结果，结合自己的论文，制定出最合理的数据呈现方式。

5.1 使用插图、表格，还是文字

选用插图、表格还是文字取决于你希望读者可以从数据中获取怎样的信息。每一种呈现形式都各有利弊。

表格的优势在于：

- 对数据的记录（无论是原始数据还是处理过的数据）；
- 对计算过程的解释，或是展现决定计算结果的各种要素；
- 对具体数值和精确度的完整呈现；
- 允许从多角度对不同要素进行比较。

图形的优势在于：

- 直观展示整体趋势；
- 通过“形状”对研究故事进行认知，而非具体数值；
- 允许少数要素间的简单比较。

图表的选用标准详见表 5.1。

表 5.1 选用图或表呈现数据

条件	选用表格	选用图形
对……进行处理时，	数字	形状
焦点为……时，	单个数据数值	整体趋势
具体或准确数值……时，	更为重要	不太重要

5.2 图形的设计

围绕你希望传达的核心信息设计每个图形。目前，你可以接触到许多制图软件，并且有能力制作大量不同风格的图形，这就更要求图形的设计为传递信息服务。在使用软件制图之前，就应该确定图形中需要包含哪些要素。这样做可以避免软件采用默认设置或现成模板生成的图形混淆重点，无法满足要求。在设计图形时，建议你考虑：

- 哪个变量需要突出显示，使用特别的符号或（加粗的）线条；
- 强调的到底是要素间的相似性还是差异性；
- 使用怎样的坐标范围、刻度间隔、极值以及统计表示法可以最有效直观地呈现数据。

使用常见的几种图形时，可以遵循下列原则：

- 饼图（pie chart）可以有效突出要素所占的比例；
- 柱状图、条形图（column and bar chart）可以有效比较相互独立、不同类别的数值（如苹果和橘子）；
- 线形图（line chart）可以呈现变量在一段时空内的变化，或其他的相关关系（如随着时间推移而改变）；
- 雷达图（radar chart）在不同类别间无法直接比较时更加有效。

你还需要在整篇论文的不同图形间保持风格一致，尤其是在不同图形中，应尽可能使用相同的符号和顺序来指代某个特定的变量或处理方法。避免图形杂乱无章，读者可能会由于图形中包含太多不同要素而无法抓住重点。

期刊在进行编辑时，可能需要压缩图形来适应版面或列宽，这或许会致使趋势曲线或符号挤在一起，难以辨认，因此也需要在设计图形时就有所考虑。试着将图形缩放到目标期刊要求的尺寸，并检查图形传递的信息是否依然清晰、明显。

要想图形在视觉上达到最理想的效果，建议满足如下条件：

- 长宽比为 3∶2；
- 图形本身不复杂时，将其置于方框内；
- 图形本身包含许多线条、条形或柱形时，不要绘制边框。

已发表的论文中使用的图形也存在许多问题，这些问题削弱了图形的直观性，从而无法有效地向读者传递信息。常见的问题如下。

- 选用了错误的图形，使得要素间的重要关系无法清晰呈现，甚至突出了某些不存在的相关性。
- 在图题中使用描述性话语，而非可以向读者提供更多信息的叙述性话语（第 10 章关于论文标题的建议在这里仍然适用）。
- 图形中重复了已经在正文或表格中呈现过的数据。
- 符号、标记或线条使用了不合理的形状、底纹或宽度，使得图形要强调的主要信息、研究结果无法得以凸显。
- 图形杂乱无章，含有太多线条、图例、数字或标出了不合理的刻度值。
- 未对坐标轴进行描述性说明，或使用特定学科分支、研究团体中难懂的术语来标记坐标轴。
- 在具体数值不太重要，或可以通过横纵坐标轴看出近似值时，依然标出数字。
- 未对数据进行分类，无法突出重点或要素间的重要关系；相关联的图形间缺乏连贯性，使读者无法对研究结果迅速进行评估。

图形细节上的微小调整可以大幅度改善信息的传递效率。以图 5.1 和图 5.2 为例，图 5.1 中已经包含了所有必要的信息，但我们在图 5.2 中做出了一些改进，使其更好地“独立存在”，更直观地突出重要信息。

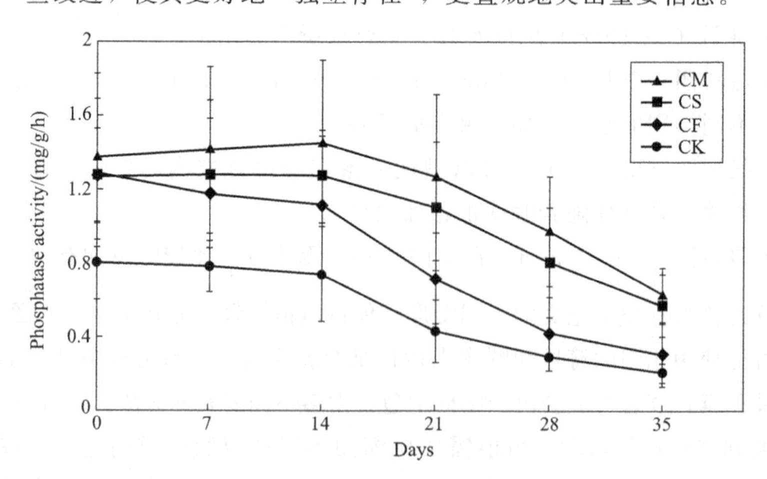

Fig. 5. 1 Comparisons of root surface phosphatase activity of wheat plants for Control (CK), exclusively chemical fertilizer (CF), combined application of chemical fertilizer and wheat straw (CS), and farmyard manure (CM) treatments. Error bars represent the standard error of the mean for each treatment.

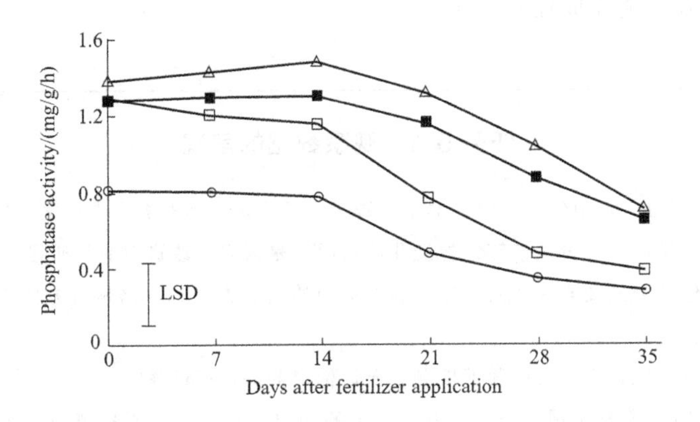

Fig. 5. 2 Root surface phosphatase activity of wheat plants differed after soil amendment with different fertilizer treatments. Phosphatase activity was highest in farmyard manure (△) treatments followed by combined application of chemical fertilizer and wheat straw (■), chemical fertilizer alone (□), and control/no amendment (○) treatments. Phosphatase activity declined over 5 weeks for all treatments. Least significant difference (LSD; two-way ANOVA, $P \leqslant 0.05$) is 0. 39mg/g/h.

与图 5.1 相比，图 5.2 进行了如下改进。

- 为降低原图的混乱程度，删掉了误差棒而是用 LSD（最低位有效数字）棒代替，这样做还可以凸显不同处理方法之间的差异，方便比较；同时调整了 y 轴坐标范围，从而扩展了 y 轴上的刻度间隔，拉开图形中曲线的间距。在图注说明中，增加了显著性差异的描述。去除了图形的边框，让画面看起来更整洁。
- 修改原 CF 组（化学肥料）和 CS 组（化学肥料加小麦秸秆）的符号为同样形状的空心、实心方形，便于对这两个主要组别进行集中比较，也更清晰地与其他组别区分开来。
- 在图注说明中对符号进行介绍，删去图例，为图形制造更多留白，让读者聚焦于曲线的差异。
- 改进了对 x 轴的描述，使其更加具体，调整了刻度间隔。
- 将图题中的文字改为叙述性话语，突出数据要传递的信息。

还有一些其他类型的图形，虽然不能用来呈现结果，但可用来描述过程（如流程图）、解释方法（如装置图）或提供书面证据，例如一些原始的影像记录（如照片或空间表征）。在使用所有的图形时，都应该遵循我们在上文中介绍的基本原则。即：

- 突出强调最重要的信息；
- 目的是为更好地呈现研究结果并提供证据；
- 图形要清晰，且在风格上保持一致，不要重复呈现已经展示过的数据。

接下来，请完成任务 5.1。

任务 5.1 观察数据的呈现

观察你的 SA 中使用了哪些类型的数据，并且是如何对数据进行呈现的。

1. 选用的呈现方式能够凸显变化趋势吗？是否可以改进得更加明显？

2. 你对哪些要素间的对比感兴趣？选用的呈现方式和风格是否方便你对要素进行比较？

3. 是否包含了足够的数据信息，允许你对数据进行计算？

4. SA 的数据呈现是否存在书中列出的明显问题？这对于描述研究结果产生了怎样的阻碍？

5.3 表格的设计

表格通常用来记录研究的数据和元数据，包含许多行或列，需要读者仔细阅读，才能确认表格要传递的信息。尤其是有时表格的单元格数量太多，且要求读者对比不同行和列之间的数据，想要迅速理解研究结果并不容易。为了最大限度克服表格在这些方面的局限性，需要对表格进行更为合理的设计，尤其是整体布局、数据选用、排列顺序、行与列的标题以及表题的文字说明。在视觉设计方面，许多原则都与图形的设计类似：避免杂乱无章，对缩写词在脚注或表题中进行定义，不要在表格外面套上边框，以及减少使用垂直线（可以采用三线表等形式）。

已发表的论文中使用的表格也存在许多问题，这些问题导致表格无法有效地向读者传递信息。常见的问题如下。

- 在表题中使用描述性话语，而非可以向读者提供更多信息的叙述性话语（第 10 章关于论文标题的建议在这里仍然适用）。
- 表格内包含了不必要的数据，例如正文未提及的数据，与研究结果无关的数据，或者重复出现的已知常量。
- 表格内包含了不够显著或过于精确的数字，给读者造成精度不够或数据杂乱的印象。
- 表格内省去了用于重要计算步骤的实验数据，在正文中也没有提供相应的信息。
- 表格内数据的排列顺序不合理，无法突出最重要的研究结果。
- 表格内的数据没有分类呈现，无法体现不同要素之间的重要关系。

表 5.2 展示的数据对比了使用不同方法分析几组土壤时，钾元素的含量。但我们在表 5.3 中做出了一些改进，使其更直观地突出相关信息。

Table 5.2 Soil test K and mineralogy of soils (SD = Standard Deviation).

Soil	Clay/(g/kg)	Silt/(g/kg)	mg K/kg soil		
			WS	$CaCl_2$	NaTPB
1	380	200	10	41	480
2	535	265	31	162	1208
3	410	230	15	57	583
4	434	205	19	70	652
5	485	235	27	100	932
6	610	282	50	290	1730
7	360	190	6	34	360
8	440	235	20	87	723
Mean	456.8	230.3	22.3	105.1	833.5
SD(±)	83.4	31.9	13.9	84.9	448.9

Table 5.3 Soil texture correlates with K concentration determined using three extraction methods: WS = Water Soluble, $CaCl_2$ = Calcium Chloride, NaTPB = Sodium Tetraphenyl Boron (SD = Standard Deviation).

Soil	Clay/(g/kg)	Silt/(g/kg)	mg K/kg soil		
			WS	$CaCl_2$	NaTPB
7	360	190	6	34	360
1	380	200	10	41	480
3	410	230	15	57	583
4	434	205	19	70	652
8	440	235	20	87	723
5	485	235	27	100	932
2	535	265	31	162	1208
6	610	282	50	290	1730
Mean	457	230	22	105	834
SD(±)	83	32	14	85	449

与表 5.2 相比，表 5.3 进行了如下改进。

- 将表题中的文字改为叙述性话语，突出数据要传递的信息。
- 将土壤样本重新排序，更好突出不同样本中粘土含量的梯度变化（表 5.2 使用的是采样顺序）。还可以对土壤样本进行重新命名，并按新的合理顺序呈现。
- 将平均数和标准差进行了取整（避免了过度精确），同时使表格更加整洁。
- 新的排序可以更好地区分单个数据点与平均值（我们甚至可以稍微增加表 5.3 中样本 6 与 Mean 之间的行间距），在观察土壤结构和钾含量的梯度时也更直观。

请完成任务 5.2。

任务 5.2 评估表格的设计

拿出 SA 或另外一篇与你研究领域相关的论文，分析其中的表格。

1. 选用的数据都是必要的吗？是否将数据进行分类以便凸显主要研究结果？

2. 表题内容是描述性文字，还是叙述性话语？能否为表格设计出叙述性的表题？

3. 表内数值有效数字的位数能否反映测量的准确程度，是否四舍五入到适当精度？

4. 表格是否存在书中列出的明显问题？这对于展现研究结果产生了怎样的阻碍？

5.4 图表说明

图注说明和表题需要对图表中呈现的数据进行解释，并突出与之相关的重要研究结果。尤其是对于当前研究中关键信息的解读，应做到“独立、完整”，读者无需阅读正文的相关内容就能够理解。清晰、有效的图表既可以提高审稿人的工作效率，也可增加文章可读性。

图注说明通常包含五个部分，按下列顺序出现，有时，第五部分“对符号或标记的说明”可能会分散在其他部分之中。

1. 图题：概述图形主要内容。

2. 对图形所呈现的研究结果或模型进行详细说明，或对图形进行补充说明。

3. 进一步解释图形中的各要素、使用的方法，或如何通过该图得出研究结果。

4. 描述图形包含的单位或统计表示法。

5. 对其他符号或标记的说明。

表题也包含类似的要素，但是第二、第三部分通常比较简要，并且第五部分也不常见。请完成任务 5.3。

任务 5.3 辨认图注说明的各个部分

下列图注说明选自书后三篇 PEA 的结果部分，请仔细阅读并辨认这些说明都包含那几个部分。

Number of *Sargassum muticum* (a) recruits and (b) adults in field experiment plots (900cm^2). Propagule pressure is grams of reproductive tissue suspended over experimental plots at beginning of experiment. The average mass of an adult *S. muticum* (174g) is indicated by an arrow. Data are means ±1 SE (n = 3). (from Britton-Simmons & Abbott 2008, Figure 1)

Uptake of Fe (II) by GmDmt1 in yeast.

(a) Influx of $^{55}Fe^{2+}$ into yeast cells transformed with GmDmt1;1, *fet3fet4* cells were transformed with GmDmt1; 1-pFL61 or pFL61 and then incubated with 1μM $^{55}FeCl_3$ (pH 5.5) for 5- and 10-min periods. Data presented are means ± SE of ^{55}Fe uptake between 5 and 10 min from three separate experiments (each performed in triplicate).

(b) Concentration dependence of ^{55}Fe influx into *fet3fet4* cells transformed with GmDmt1;1-pFL61 or pFL61. Data presented are means ± SE of ^{55}Fe uptake to over 5 min (n = 3). The curve was obtained by direct fit to the Michaelis-Menten equation. Estimated K_M and V_{MAX} for GmDmt1;1 were (6.4 ± 1.1)μM Fe(III) and (0.72 ± 0.08)nM Fe(III)/min//mg protein, respectively.

(c) Effect of other divalent cations on uptake of $^{55}Fe^{2+}$ into *fet3fet4* cells transformed with pFL61-GmDMT1;1. Data presented are means ± SE of ^{55}Fe (10μM) uptake over 10 min in the presence and absence of 100μM unlabeled Fe^{2+}, Cu^{2+}, Zn^{2+} and Mn^{2+}. (from Kaiser et al. 2003, Figure 5)

A sequence of SEVIRI images recorded from 9:15 to 10:30 GMT on 13 May. At 9:15 GMT, no evident eruptive phenomena are observable. At 9:30 GMT the image shows the beginning of an ash plume (yellow pixels) moving toward NE, which was associated to the lava fountain from the northern part of the eruptive fissure. The first hot spot, due to the increased lava output and thermal anomaly, was detected by HOTSAT at 10:30 GMT. (from Ganci et al. 2012, Figure 4)

请核对书后的参考答案。

6. 描述研究结果

论文的结果部分是体现当前研究新颖性和重要意义的绝佳位置，从中应能得出论文对当前研究领域的贡献。期刊编辑和审稿人会仔细评估研究结果的价值、影响力，并判断其是否与期刊的目标一致。结果部分汇集了视觉、数字和文字信息，将读者的注意力集中到当前研究的优势上。

在结果部分，好的作者只会强调要点。期刊编辑和有经验的作者还会建议：论文作者无需在正文中重复描述图表信息。这也就意味着结果正文中的语句应该集中陈述最重要的研究发现，尤其是那些将在讨论部分使用的要点。

6.1 结果部分的结构

结果与讨论有时是独立的两个部分，有时会结合在一起；在 AIBC 结构中，结果可能独立，也可能与方法和讨论共同形成论文正文（Body sections）的一个部分。在写作前，应查阅目标期刊的“投稿说明”，了解相关要求；如果没有明确的要求，还应该翻阅期刊中已经发表的一些论文，分析文章结构。

如果结果部分和讨论部分独立存在，写作时应注意不要在结果部分对数据进行过多分析、解释，也不要与其他研究进行比较，仅对研究结果（数据）进行描述即可。然而，有的作者还是会选择在结果部分与其他研究者的工作进行对比，可能的原因是：相关信息不会在讨论部分详细说明，但作者仍希望对某一发现进行对比和解释。例如：书后 PEA 1 中结果部分“Protein localisation”小标题下，“For example”后面的相关内容。

一般来说，结果和讨论经常是分开的。在这一章，我们也仅学习结果正文中句子的写作。

6.2 结果正文中句子的功能

结果正文中的语句（以及结果与讨论结合在一起时，描述结果的句子）通常：

- 强调重要的研究发现；
- 向读者指示在何处可以找到描述相关结果的图、表；
- 可能会对结果进行评论。

强调研究发现（highlight）和说明图表位置（location）的语句，可以连成一句，也可拆分成不同的句子。

下面是“highlight + location”连成一句的例子：

Measurements of root length density (Figure 3) revealed that the majority of roots of both cultivars were found in the upper substrate layers.

The response of lucerne root growth to manganese rate and depth treatments was similar to that of shoots (Figure 2).

下面是独立使用一句话，对图表位置进行说明的例子：

Figure 17 shows the average number of visits per bird.

注意上述两种情况下，动词时态的不同。请完成任务 6.1。

任务 6.1 找出独立说明图表位置的语句

翻开你选中的 PEA，快速阅读其结果部分。找出独立成句，对图表位置进行说明的情况，统计句子数量。你认为论文作者为何选用这种策略来呈现结果部分？请核对书后的参考答案。

翻开 SA，做同样的练习。如果可以的话，与同事讨论你的发现。

6.3 结果部分正文中动词的时态

请先完成任务 6.2。

任务 6.2 结果部分的动词使用

1. 阅读下面 PEA1 中结果部分的选段，分析划线处动词使用的时态或形式（现在时、过去时还是情态动词），并思考作者使用不同时态的原因（文中仅标出了谓语动词，未考虑过去分词用作形容词的情况）。

Antibodies were raised in rabbits against the N-terminal 73 amino acids of GmDmt1;1 (Figure 1c). This antiserum was used in Western blot analysis of 4-week-old total soluble nodule proteins, nodule microsomes, PBS prot eins and PBM, isolated from purified symbiosomes. The anti GmDMT1 antiserum identified a 67-kDa protein on the PBM-enriched nodule protein fraction (Figure 3a), but did not cross-react with soluble nodule proteins, PBS proteins or nodule microsomes (Figure 3a). Replicate Western blots incubated with preimmune serum (Figure 3b) did not cross-react with the soybean nodule tissue examined. The protein identified on the PBM-enriched protein fraction is approximately 10 kDa larger than that predicted by the amino acid sequence of GmDmt1. The increase in size may be related to extensive post-translational modification (e.g. glycosylation) of GmDmt1, as it occurs in other systems. (Kaiser et al. 2003)

2. 根据你的分析结果，补全下列句子：

- In Results sections, the past tense is used to talk about ...
- The present tense is used in sentences that ...
- Modal verbs are used to ...

将你的答案与下面“结果部分的常用时态”相关要点进行对比。

结果部分的常用时态

- 对于当前研究进行描述时，使用一般过去时（主动或被动语态），说明做了什么、发现了什么。
- 现在时用于如下几种情况。
  - 永远正确的“真理性”描述。
  - 介绍某份文件内容时，会将文件视为一种客观存在。任务 6.2 节选的 PEA 1 中的段落不涉及这种用法，请看下面的例子（McNeill et al. 1997）：

    The effect of urea concentration on the fed leaf and shoot growth in subterranean clover is summarised in Table 1.
- 对结果进行评论时，可能会使用情态动词（例如 may 和 could），在 that 引导的从句中尤其常见，详见第 9 章有关情态动词的说明。

请完成任务 6.3，巩固相关知识。

任务 6.3 分析 SA 结果正文部分动词的使用

阅读 SA 结果正文中的一个部分，然后回答下列问题：

- 对于每一个动词，你认为作者为何选用那样的时态？
- 作者的时态选用是否符合上文总结的规律？如有特例，你能否说出原因？

如果你发现许多不符合上述规则的情况，建议你再阅读 2～3 篇自己研究领域的论文，查看它们结果部分动词时态的选用。如果能找出大量特例，这意味着你发现了新的规则。请详细记录你的发现，并在自己的写作中参考这些规则。

提示：从自己研究领域的论文中找出的规律，对自己的写作最有指导价值。

本书不可能概括所有研究领域论文的写作规范，因此，我们的目的不在于给出所有答案，而是提示你该使用怎样的方式分析论文范例，从哪些角度去思考论文的语言。每当你读到本书列出的相关规则时，都请核对它们是否适用于自己领域的论文，并据此对这些规则进行调整，以便于你在未来写作时，准确把握写作方法，达到目标期刊的要求。就像我们在开头提到的：本书只负责“描述”特征，而不是“规定”你该如何去做。将你的自选论文与我们描述出来的特征进行对比，得到的结论才是对你而言最有价值的信息。如果你已经准备好了研究结果，请继续完成任务 6.4。

任务 6.4 开始 OA 结果部分的写作

请完成论文初稿（OA）结果部分的写作，并在做好图表设计后，完善相关文字内容。

7. 方法部分

7.1 方法部分的作用

学生受到的传统教育总是在讲，方法部分可以为其他有条件的研究者提供信息，重复论文所介绍的研究活动。结合你阅读论文的感受，你是否同意这一观点？参与我们“论文写作研讨会”的许多研究人员都表示，即使仔细阅读了论文的方法部分，在重复作者介绍的实验时仍然困难重重。

解读这个问题的另一个角度则是：方法部分应该为证明研究结果的信度提供足够信息，读者可以通过这些信息对结果进行评估，并确定它们是否和作者所声称的研究发现相一致。同时，审稿人们在阅读方法部分时，也会不断寻找证据来评估论文作者的方法和对结果的处理是否符合科学规范。

论文的方法部分可能有不同的命名方式。在第 2 章分析 PEA 结构时，你应该已经发现了这一点，方法部分的标题可以是 Methods，Materials and methods，或 Experimental procedures。如果论文是 AIBC 结构，方法可能会分布在正文不同的部分中，并且不会直接使用“method”这个单词作为小标题。本章会统一使用“方法部分”这一简洁的名称。

如果参考了前人公开发表过的实验方法，通常可以进行引用，省去详细说明，除非对原有步骤作了修改。有时，国际读者不方便获取相关的参考文献，亦或许原始文献的语言不是英语，建议你在引用文献的同时，仍然对研究方法进行详细介绍；如果可能的话，在对应的参考文献中也使用括号补充说明文献的语种。如果你的研究使用了全新的方法，务必要完整描述，以确保审稿人可以获取足够的信息，并判断研究方法是否适当、有效。

7.2 方法部分的呈现

读者应该能通过阅读方法部分，对结果部分的研究发现进行评估。对于论文的作者而言，有必要在上述两个部分之间建立联系。还应注意的一点是，读者可能在阅读论文任何其他部分时，直接跳跃到方法部分查找相关内容。德国学者 Joy Burrough-Boenisch 研究发现（1999），参与调查的期刊编辑和审稿人中，29％～33％会略过方法部分随后阅读，而 21％～27％则会选择更早阅读。为了更清晰地阐明方法部分和结果部分的联系，本书推荐两种策略。

- 策略 1：在方法部分和结果部分，使用相似甚至同样的小标题。
- 策略 2：在方法部分的语句中，加入介绍性的语句，说明进行某项操作要达到的目标。例如：

To generate an antibody to GmDmt1;1, a 236-bp DNA fragment coding for 70 N-terminal amino acids was amplified using the PCR, ...

还有一种策略，就是在新段落的开头承上启下：在第一句话中表达你即将阐述的主题，并与之前叙述过的内容有效衔接。在下面的例子（见PEA2）中，"disturbance treatment"是上文已经介绍过的概念，而句子的主干也进一步向读者说明了后文要涉及的主要内容。

The disturbance treatment had two levels: control and disturbed. Control plots were ... (Britton-Simmons & Abbott 2008)

许多研究领域对于方法部分如何选用小标题以及如何排序都有标准的要求，建议你仔细查阅目标期刊的论文，采用相似的方式呈现自己的研究方法。小标题应该是越详细越好，读者在心中存有疑惑时能迅速定位到答案，理解研究是如何实施的。请完成任务 7.1 和任务 7.2。

任务 7.1 "材料和方法"部分的呈现

翻开你选中的 PEA（如果你的选择是 PEA3，请阅读论文的第 2 部分），回答如下问题。

1. 方法部分使用了哪些小标题？
2. 这些小标题与引言部分的结尾和结果部分的小标题分别有何联系？
3. 论文的方法部分是否读者友好？有什么样的表现？如果在阅读时遇到困难，原因是什么？

请核对书后的参考答案。

翻开 SA，做同样的练习。如果可以的话，与同事或老师讨论你的发现。

任务 7.2 开始 OA 方法部分的写作

你希望论文初稿的方法部分包含哪些要素？以何种顺序呈现？

7.3 使用被动语态和主动语态

研究者在写作方法部分时会使用大量被动语态。这样做可以引导读者关注动作本身，而非动作的执行者，但是与主动形式相比，被动语态会导致用词数量的增多。你也会读到许多建议，要求你尽量使用主动语态，这可以避免冗长的语言，使之更加直接。在我们看来，科技文体中，确实存在过度使用被动语态的现象。可是动词的语态也并非通过简单的选择就能确定，尤其是在论文的方法部分。接下来，本书将与你一起：

- 回顾被动语态与主动语态的差异；
- 思考写作者有意使用被动语态的原因；
- 理解并解决使用被动语态时经常出现的问题。

被动语态与主动语态的形式差异

在使用主动语态时，动词的主语实际上是动作的发出者。例如：

subject	+	active verb	+	object
The dog		**bit**		**the man.**

在被动语态中，句子的主语不是动作的发出者。在下面的例子中，发出“biting”动作的不是“the man”，即：

subject	+	passive verb	+	agent
The man		**was bitten**		**by the dog.**

在使用被动语态的句子中，动作的发出者经常省略，因此，当我们需要强调动作本身的时候，也就是在描述实验流程时，更倾向于使用被动语态。图 7.1 概括了两种句子结构的差异。

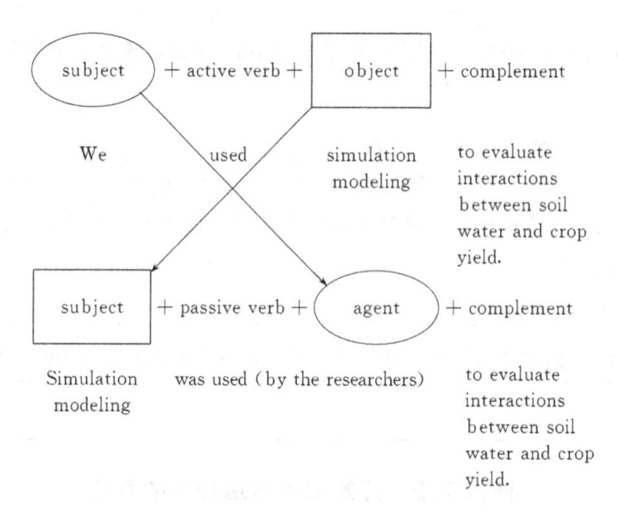

图 7.1 主动句变为被动句

如果作者本人不介意使用主动语态，像图 7.1 中那样使用“we”作为句子的主语，那么在写作中避免被动语态也就相对更加容易，即便是在方法部分。但是，许多作者不喜欢这种用法，尤其是每一句都以“we”开头时，这样的重复降低了阅读的愉悦感，因此在科技文体中，被动语态的使用频率依然很高。

在构成被动语态时，需要添加助动词“be”的某个形式（比如在例句中使用的 was），再加上动词的过去分词（比如在例句中使用的 bitten）。注意，只有及物动词（词典中标记为 vt. 的动词）才可以转化为被动句，因为不及物动词后面没有宾语。请完成任务 7.3。

任务 7.3 主动句和被动句

翻开你选中的 PEA，从方法部分找出一个被动句，将它转化为主动句。再找出一个使用及物动词的主动句，将它转化为被动句。我们在每篇 PEA 中都选取了两句话进行转换，请核对书后的参考答案。

被动语态与主动语态的选用依据

首先思考，读者是否有必要了解动作的发出者。如果这一信息不重要，你就可以使用被动语态。请看下面的例子：

The researchers collected data from all sites weekly.

在这个语境中，谁来采集数据并不重要，因此可以使用被动句：

Data were collected weekly from all sites.（注意：data 是复数形式，源自拉丁语。许多编辑仍然对 data 后面使用的动词形式有强制要求。但是这一规则也在变化之中，你在阅读文献时，或许既能看到 the data show，也能看到 the data shows）

其次，频繁使用人称代词做主语是否读起来有些重复，或显得作者不够谦虚？例如：

We calculated least significant differences (l. s. d.) to compare means.

使用被动句或许更为恰当，即：

Least significant differences (l. s. d.) were calculated to compare means.

在选择动词的语态时，应参考下列因素。

- 在 PEA1 的方法部分（Experimental procedures），几乎没有使用主动句，主要原因应该是为了避免重复。如果使用主动句，每句话可能都要用“we”开头。
- 如果在你的学科中，独自撰写论文的情况很常见，建议你多翻阅已经发表的论文，判断使用“I”是否合适。以我们的经验，在科技文体中，尤其是在方法部分，这样的写法并不常见。
- 在语境中，对比主动语态和被动语态，看哪一种更有助于信息流动。

使用英语进行写作时，好的作者总能在语句之间建立有效的联系，主要表现在新、旧信息的合理排布上（详见第 8 章 8.9 节）。有时，使用被动语态可以帮我们达到上述目的。在下面的例子（见 PEA2）中，旧信息使用斜体标出，同时用方括号备注了语态选用情况。

We used [active] the results of these analyses to inform the construction of mechanistic candidate functions for the relationship between propagule input, space availability and recruitment. *These candidate functions* were compared [passive] using differences in the Akaike information criteria (AIC differences; Burnham and Anderson 2002). We then used model averaging [active] ... (Britton-Simmons & Abbott 2008)

被动句常见问题

最容易出现的问题是句子太过笨拙，不利于读者理解，主要表现在头重脚轻。因此强烈建议你不要把过长的主语置于句首，并且把“be”动词以及过去分词放在句子末尾，例如：

× Wheat and barley, collected from the Virginia field site, as well as sorghum and millet, collected at Loxton, were used.

修改思路是：尽量让读者在句子的前九个词之内就读到句子的主语和谓语动词，将罗列式的信息置于句子后部。

√ Four cereals were used: wheat and barley, collected from the Virginia field site; and sorghum and millet, collected at Loxton.

注意上面修改后的句子，这个句式适用于罗列信息。先使用一个分句（或者整句话）介绍主要内容，然后使用冒号引出后面的列表。如果同一个类别的信息内部使用了逗号，在区分不同类别时，可以使用分号。借助标点符号，可以令读者对相关信息一目了然。请完成任务 7.4。

任务 7.4　头重脚轻的被动句

1. 修改下面头重脚轻的句子，由于是被动句，很长的主语放置在句首，导致读者无法迅速抓住句子的主要信息。

Actual evapotranspiration (T) for each crop, defined as the amount of precipitation for the period between sowing and harvesting the particular crop plus or minus the change in soil water storage in the 2m soil profile, was computed by the soil water balance equation (Xin 1986; Zhu and Niu 1987).

From Li et al. (2000).

请核对书后的参考答案。

2. 阅读 SA 方法正文中的一个部分，查找是否包含头重脚轻的被动句。有哪些语句不利于读者理解？能否改进？与同事讨论你的发现。

简化被动句，避免重复

表 7.1 提供了一些简化被动句的方法。请在学习后，完成任务 7.5。

表 7.1 简化被动句，避免重复

原句	简化后
The data were collected and they were analysed using ...	The data were collected and analysed using ...
The data were collected and correlations were calculated ...	The data were collected and correlations calculated ...
The data which were collected at Site 1 were analysed using ...	The data collected at Site 1 were analysed using ...①

① 在没有以下状语的情况下（如“at Site 1”），该句的常见表述为“The collected data were analysed”。

任务 7.5 修改 OA 的方法部分

请结合本章内容，对论文初稿的方法部分进行修改。

8. 引言部分

正如第 1 章所述，期刊编辑和审稿人是在你投出稿件后的首批读者，而他们阅读的起点很可能就是引言部分，因此这部分的写作非常重要。审稿人会通过阅读引言，寻找下列问题的答案。

1. 论文的研究问题新颖吗？
2. 论文的研究内容是否有重要意义？
3. 是否适合在本刊发表？

8.1 层层深入，增加论述的说服力

应用语言学的研究结果表明，引言部分在展开论述时，可以分为不同的层级（图 8.1）。研究者们分析了许多已经发表的论文，确立了六个主要的层级，在不同的学科中，有的层级会包含不同的内容。但是就本书而言，这些层级所构成的框架足够我们针对不同学科的文本进行分析。但是也请不要不假思索地直接套用这些结构；相反，我们希望你通过了解引言的框架，针对自己研究领域的论文情况进行分析，用分析、对比的结果引导自己的写作。

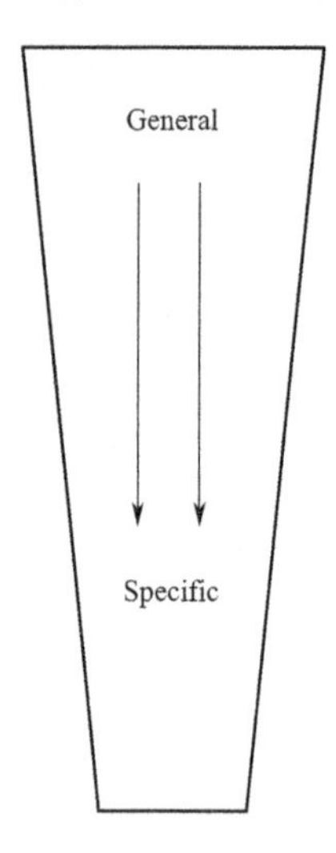

（1）陈述当前研究领域，为读者介绍研究问题所处的背景，并确立其重要性或核心地位。
（2）针对研究问题，陈述其他研究者的研究角度和已经开展的研究工作。向读者提供“已知信息”。
（3）指出进一步研究的必要性，告知读者存在研究空白。为提出当前研究活动做好铺垫。
（4）具体说明当前研究的目标，或概述主要的研究活动、研究发现。
（5）陈述当前研究的意义和价值（非必备层级）。
（6）说明文章构思，预告论文框架（非必备层级，取决于特定的研究领域）。

图 8.1 科技论文引言部分的层级（基于 Weissberg & Buker 1990）

引言部分的这些层级有时不会按照图 8.1 中的顺序出现，有的层级可能会在引言中出现多次。例如，当作者从不同的方面或角度解释研究的必要性时，“层级 2+层级 3”这种组合可能会反复出现。为了帮助你理解这些层级的内容，请首先阅读表 8.1 中的论文引言，并思考我们对于层级的划分。然后完成任务 8.1。

表 8.1 引言部分的层级划分

"Use of *in situ* ^{15}N-labelling to estimate the total below-ground nitrogen of pasture legumes in intact soil-plant systems" (McNeill et al. 1997)

引言部分	层级
Current estimates of the below-ground production of N by pasture legumes are scarce and rely mainly on data from harvested macro-roots (Burton 1976; Reeves 1984) with little account taken of fine root material or soluble root N leached by root washing. Sampling to obtain the entire root biomass is extremely difficult (Sauerbeck and Johnen 1977) since many roots, particularly those of pasture species (Ellis and Barnes 1973), are fragile and too fine to be recovered by wet sieving. Furthermore, the interface between the root and the soil is not easy to determine and legume derived N will exist not only as live intact root but in a variety of other forms, often termed rhizodeposits (Whipps 1990).	层级 1 层级 3 (scare; little account) 层级 1
An approach is accordingly required which enables *in situ* labelling of N in the legume root system under undisturbed conditions coupled with subsequent recovery and measurement of that legume N in all of the inter-related below-ground fractions. Sophisticated techniques exist to label roots with ^{15}N via exposure of shoots to an atmosphere containing labelled NH_3 (Porter et al. 1972; Janzen and Bruinsma 1989) but such techniques would not be suitable for labelling a pasture legume within a mixed sward.	层级 3(broad gap) 层级 2 层级 3
Labelled N_2 atmospheres (Warembourg et al. 1982; McNeill et al. 1994) have been used to label specifically the legume component of a mixed sward via N_2 fixation in nodules.	层级 2
However, these techniques require complex and expensive enclosure equipment, which limits replication and cannot be easily applied to field situations; furthermore, non-symbiotic N_2 fixation of label may occur in some soils and complicate the interpretation of fate of below-ground legume N.	层级 3
The split-root technique has also been used to introduce ^{15}N directly into plants by exposing one isolated portion of the root system to ^{15}N either in solution or soil (Sawatsky and Soper 1991; Jensen 1996), but this necessitates some degree of disturbance of the natural system. Foliar feeding does not disturb the system and has the additional advantage that shoots tolerate higher concentrations of N than roots (Wittwer et al. 1963). Spray application of ^{15}N-labelled urea has been successfully used to label legumes *in situ* under field conditions (Zebarth et al. 1991) but runoff of ^{15}N-labelled solutions from foliage to the soil will complicate interpretation of root-soil dynamics. Russell and Fillery (1996), using a stem-feeding technique, have shown that *in situ* ^{15}N-labelling of lupin plants growing in soil cores enabled total below-ground N to be estimated under relatively undisturbed conditions, but they indicated that the technique was not adaptable to all plants, particularly pasture species. Feeding of individual leaves with a solution containing ^{15}N is a technique that has been widely used for physiological studies in wheat (Palta et al. 1991) and legumes (Oghoghorie and Pate 1972; Pate 1973). The potential of the technique for investigating soilplant N dynamics was noted as long as 10 years ago by Ledgard et al. (1985) following the use of ^{15}N leaf-feeding in a study of N transfer from legume to associated grass. The experiments reported here were designed (1) to assess the use of a simple ^{15}N leaf-feeding technique specifically to label *in situ* the roots of subterranean clover and serradella growing in soil, and (2) to obtain quantitative estimates of total below-ground N accretion by these pasture legumes.	层级 2 层级 3 层级 2 层级 3 层级 2 层级 3 层级 2(在 potential 部分暗含层级 3) 层级 4(当前研究的目标)

任务 8.1 引言部分的层级

翻开你选中的 PEA，阅读引言部分，分析出现了哪些层级，并把它们标注出来（注意：有些层级可能重复出现；出现的顺序有时也跟图 8.1 所示的顺序不同）。

请核对书后的参考答案。

翻开 SA，做同样的练习。如果可以的话，与同事或老师讨论你的发现。

8.2 层级 1：在当前科研领域中定位出你的研究项目

合理构建论文的研究背景

在层级 1 的开始，作者通常会概述读者们已经广泛接受的事实。在使用英语进行“真理性”描述时，选用的时态通常为一般现在时；常用的时态还有现在完成时，用于表达站在当前时间点对过去已经完成的研究进行回顾。在进行这部分陈述时，使不使用参考文献都是允许的，这取决于具体的研究领域和论文的主题。请完成任务 8.2。

任务 8.2 分析引言部分的层级 1

1. 翻开书后的三篇 PEA，对引言部分的第一段进行分析，然后完成表 8.2。请核对书后的参考答案。

2. 翻开 SA，做同样的练习。将分析结果与表 8.2 中整理出的信息进行对比。如果可以的话，与同事或老师讨论你的发现。

表 8.2 任务 8.2：分析引言部分的层级 1

问 题	PEA1	PEA2	PEA3
是否有使用一般现在时的句子？有几句？			
是否有使用现在完成时的句子？有几句？			
哪种时态使用更多？可能的原因是什么？			
有几句话使用了参考文献？			
哪类句子没有引用参考文献？			

从宏观上概述过研究领域后，作者通常会把叙述范围缩小到某个子领域，最后顺利过渡到自己的研究主题上。可以这样想：引言的开头像是在选定“国家”，你从“国家”这个广阔领域开始移动，进一步定位到其中的一个“省份”，最后聚焦到特定的那座“城市”，“城市”代表你研究活动所属的研究主题。请完成任务 8.3，理解层级 1 的写作思路。

任务 8.3　层级 1：从"国家"到"城市"

1. 翻开你选中的 PEA，对引言部分进行分析，你能否找出其中的"国家"、"省份"和"城市"对应的内容？

请核对书后的参考答案。

2. 翻开 SA，做同样的练习。指出引言部分的"国家"、"省份"和"城市"。

3. 针对你论文初稿（OA）的引言部分，思考"国家"、"省份"和"城市"这三个特征。注意："城市"代表的是你论文所属的话题领域，而非你的研究目的。

作者带领读者在上述内容间移动时，需要合理的布置新旧信息。旧的信息代表读者已知的信息，应放置在句子的开头，新的信息则应放置在句子的后部（在科技文体中，这样做可以更好地促进信息流动，使文章更加连贯）。请完成任务 8.4。

任务 8.4　辨识旧的信息

下面的段落选自 PEA1 的引言部分，请你一边阅读，一边标出已知信息（在代表旧的信息的词语下划线）。

Legumes form symbiotic associations with N_2-fixing soil-borne bacteria of the *Rhizobium* family. The symbiosis begins when compatible bacteria invade legume root hairs, signalling the division of inner cortical root cells and the formation of a nodule. Invading bacteria migrate to the developing nodule by way of an 'infection thread', comprised of an invaginated cell wall. In the inner cortex, bacteria are released into the cell cytosol, enveloped in a modified plasma membrane (the peribacteroid membrane (PBM)), to form an organelle-like structure called the symbiosome, which consists of bacteroid, PBM and the intervening peribacteroid space (PBS; Whitehead and Day, 1997). The bacteria, subsequently, differentiate into the N_2-fixing bacteroid form. The symbiosis allows the access of legumes to atmospheric N_2, which is reduced to NH_4^+ by the bacteroid enzyme nitrogenase. In exchange for reduced N, the plant provides carbon to the nodules to support bacterial respiration, a low-oxygen environment in the nodule suitable for bacteroid nitrogenase activity, and all the essential nutritional elements necessary for bacteroid activity. Consequently, nutrient transport across the PBM is an important control mechanism in the promotion and regulation of the symbiosis.

请核对书后的参考答案。

8.3　在层级 2、层级 3 中使用参考文献

在引言的层级 2 和层级 3（图 8.1）中，作者通过使用研究领域的相关文献，证明自己研究工作的合理性，并借助文献指出进一步研究的必要性，为提出自己的研究活动做好铺垫。这部分内容需要作者引用自己选出的文

献，即特定研究领域内已经发表的研究论文、综述和图书，此外，参考文献也可来自发布在网站上的内容，但这些内容需要经过同行评议或属于权威科学机构。

如何引用参考文献?为什么要使用参考文献?

结合上面给出的案例，你应该发现在引言的各个层级，都允许引用参考文献（references，citations 或 in-text citations)。文献在文中可能以姓氏、出版年份加括号的形式出现，例如“(McNeill 2000)”；也可能使用数字形式，例如“[7]”。期刊会对此进行详细的规定，建议你仔细阅读目标期刊“投稿说明”中的相关内容。正文中的文献都指向论文最后的参考文献列表，此时则需要逐条列出详细的出版信息，具体的呈现格式也应参考“投稿说明”。

请注意，你所使用的参考文献代表你是否了解当前话题领域（任务 8.3 中的“城市”）的其他研究者已经做了哪些工作，据此，你才能提出还有哪些未做的工作亟待完成，而你的研究工作即将填补相关的研究空白。这也是在层级 2 和层级 3 中需要完成的主要任务。你事实上是通过组织参考文献，论证自己研究的合理性，并说明自己的研究为何重要。

利用参考文献，展开自己的论证

下面列举三种利用参考文献的方法（相关文献仅用于举例，并非真实存在)。这些方法可以称为：信息凸显（information prominent)，语句的核心是呈现信息；作者凸显（author prominent)，在语句中强调研究者的名字；弱作者凸显（weak author prominent)，语句涉及“作者”这一概念，但不会强调具体某位研究者。上述形式都可用于帮助论文作者展开论证，进一步观察下面的例子，体会不同形式适用的写作目的。(本段中，author 指的是参考文献的“作者”，请注意与“论文作者”进行区分)

信息凸显

Shrinking markets are also evident in other areas. The wool industry is experiencing difficulties related to falling demand worldwide since the development of high-quality synthetic fibres (Smith 2000) .

在许多科学领域，上述方法都属于默认的引用形式。书后有两篇 PEA 的引言甚至通篇都使用了信息凸显的方法。此外，还有两种引用形式供论文作者选用。

注意：上例中的第一句话是段落的主题句，由于上文也讨论了 shrinking markets，这句话可以用来构成衔接，同时向读者说明本段的主要内容。在科技文体中使用主题句，可以理顺前后文的逻辑。

作者凸显 1

Shrinking markets are also evident in other areas. As Smith (2000) pointed out, the wool industry is experiencing difficulties related to falling demand worldwide since the development of high-quality synthetic fibres.

这种形式允许论文作者表明立场。上例中，论文作者或者说你本人，认同 Smith 的观点。

作者凸显 2

Shrinking markets are also evident in other areas. Smith (2000) argued that the wool industry was experiencing difficulties related to falling demand worldwide since the development of high-quality synthetic fibres. However, Jones et al. (2004) found that industry difficulties were more related to quality of supply than to demand issues. It is clear that considerable disagreement exists about the underlying sources of these problems.

这种形式在使用“argued”的同时，间接提醒读者后文中可能会出现“however”或其他对立的观点，也说明论文作者（你本人）未必认同正在引用的这篇文献。请注意，如果过度使用作者凸显的形式，文章读起来会更像在罗列不同研究者的观点，无法构成有效的论证。建议你严格控制作者凸显在引言中使用的次数，如果有必要，在即将指出研究空白时，可以偶尔使用。请看 PEA 3 引言部分使用作者凸显 2 的实例。

These modeling efforts all used infrared satellite data acquired by AVHRR and/or MODIS to estimate the time-averaged discharge rate (TADR) following the methodology of Harris et al. (1997) . Recently, we developed...

在决定如何使用作者凸显的形式时，应仔细阅读你研究领域中已发表的论文，查看这种引用方法的使用频率。

弱作者凸显

Several authors have reported that the wool industry is experiencing difficulties related to falling demand since the development of high-quality synthetic fibres (Smith 2000, Wilson 2003, Nguyen 2005) . For example, Smith (2000) highlighted...

这种形式通常使用“authors”做主语，然后在括号中列出多个参考文献。在上例中，第二句话紧接着使用了作者凸显的引用形式，第一句话则用作段落的主题句。这样的段落构成，适用于另起一段开始介绍某个新的子话题或展开另一层论证。弱作者凸显通常会使用现在完成时（have reported）。

论文作者根据展开论证的需要，选择恰当的引用方法。请完成任务 8.5。

任务 8.5 案例分析

表 8.3 中的段落节选自 McNeill et al.（1997）的引言部分，请你一边阅读，一边观察不同引用形式的使用。

表 8.3 如何利用参考文献展开论证

引言部分	引用形式
Foliar feeding does not disturb the system and has the additional advantage that shoots tolerate higher concentrations of N than roots (Wittwer et al. 1963).	信息凸显
Spray application of ^{15}N-labelled urea has been successfully used to label legumes *in situ* under field conditions (Zebarth et al. 1991)	信息凸显
but runoff of ^{15}N-labelled solutions from foliage to the soil will complicate interpretation of root-soil dynamics.	论文作者的评价
Russell and Fillery (1996), using a stem-feeding technique, have shown that *in situ* ^{15}N-labelling of lupin plants growing in soil cores enabled total below-ground N to be estimated under relatively undisturbed conditions, but they indicated that the technique was not adaptable to all plants, particularly pasture species.	作者凸显
Feeding of individual leaves with a solution containing ^{15}N is a technique that has been widely used for physiological studies in wheat (Palta et al. 1991) and legumes (Oghoghorie and Pate 1972; Pate 1973).	信息凸显
The potential of the technique for investigating soil-plant N dynamics was noted as long as 10 years ago by Ledgard et al. (1985) following the use of ^{15}N leaf-feeding in a study of N transfer from legume to associated grass.	作者凸显，句子使用了被动语态，将 technique 置于主语的位置，属于上文介绍的旧信息，与下文构成衔接。

无法获得原始文献时的引用方法

期刊编辑会要求论文作者引用亲自读过的文献。但有时你无法获取原始文献，只能依靠其他作者对该文献的解读，在转引时，可在正文中参考下面的形式：

[The finding or fact you want to cite] (Smith 1962, cited in Jones 2002).

对于上例，参考文献列表中只出现 Jones (2002)。

8.4 引用时避免剽窃

引用时，还可能在无意中构成剽窃。剽窃是指未标明来源就使用了其他研究者的数据、想法或文字，属于学术界或出版界的欺诈行为。如果论文存在剽窃问题，会遭到拒稿。引用不完整，说明论文作者对当前研究领域的理解不够深入，很难取得其他研究者的信任。有效且完整的引用文献，可以体现出你对研究现状的了解，也容易给审稿人留下好的印象。读者在阅读论文时，也能从你使用的参考文献中有所收获。

要避免剽窃，论文作者应首先了解何时可能会由于一时疏忽，导致剽窃的发生。同时，还应学会正确地做笔记，将笔记涉及的文献内容整理到论文中时，要明确信息来源。请记住，读者有必要清楚地知道论文每句话所代表的事实或观点，到底来自于你本人，还是其他研究者。如果是后者，请务必进行引用，因为最先提出这些事实或观点的研究者很可能会成为你论文的审稿人。无论如何，审稿人对当前研究领域的文献一定相当了解，在引用时，务必要诚实、准确。请完成任务 8.6。

任务 8.6　剽窃的鉴别

下面的两段文本改编自 McNeill et al.（1997）的引言。两个版本包含的信息相同，但 Version 2 中，论文作者试图把其他研究者的观点写成自己的，请对比 Version 1 进行阅读，找出涉嫌剽窃的部分。

Version 1 Russell and Fillery（1996），using a stem-feeding technique，have shown that *in situ* ^{15}N-labelling of lupin plants growing in soil cores enabled total belowground N to be estimated under relatively undisturbed conditions，but they indicated that the technique was not adaptable to all plants，particularly pasture species.

Version 2 Russell and Fillery（1996），using a stem-feeding technique，have shown that *in situ* ^{15}N-labelling of lupin plants growing in soil cores enabled total belowground N to be estimated under relatively undisturbed conditions. However，this technique is not adaptable to all plants，particularly pasture species.

请核对书后的参考答案。

在科技文体中，加双引号进行直接引用并不常见。论文作者有必要在引用时进行转述（paraphrase），而非照抄原文。但这不影响你向他人论文中使用的精彩语句学习，你可以提炼句子模板，删去原句中的“内容区段”（通常是名词短语，在下面的例句中用 NP 表示），保留句子的框架，然后在自己的论文中加入新的内容区段使用。任务 8.6 中 Version 1 的句子就可以提炼出下面的框架：

[Authors]，using [NP1]，have shown that [NP2] enabled [NP3] to be estimated under [adjective] conditions，but they indicated that the technique was not adaptable to all [NP4]，particularly [NP5].

你在写作时，就可以利用上面的框架。有关句子模板提炼的详细介绍，请参考第 17 章相关内容。

8.5 指出研究空白

在引言的层级 3（图 8.1）中，作者会采用多种形式指出研究空白。在前面的例子中，作者会在引言的开头先提出比较宽泛的研究需要，然后在接近结尾处将范围缩小到具体的研究空白上。下面的几个例子选自 PEA 2 的引言部分。

However, understanding how these processes interact to regulate invasions remains a major challenge in ecology.

Despite its acknowledged importance, propagule pressure has rarely been manipulated experimentally and the interaction of propagule pressure with other processes that regulate invasion success is not well understood.

It is presently unclear how different disturbance agents influence long-term patterns of invasion.

在层级 3 中，经常需要用到一些信号词，例如上面句子中的 "however, remains a major challenge, rarely, not well understood" 以及 "presently unclear"。请完成任务 8.7 和任务 8.8。

任务 8.7 指出研究空白: 信号词

重新阅读表 8.1 中 McNeill et al.（1997）的引言，并翻开你选中的 PEA，阅读引言部分，标出论文作者指出研究空白时使用的信号词。将这些信号词单独列出，并核对书后的参考答案。

任务 8.8 开始 OA 引言部分层级 3 的写作

请完成论文初稿（OA）引言部分层级 3 的写作。

8.6 层级 4: 陈述研究目的

引言部分结束前，要告知读者可以期待从论文后面的部分读到哪些内容，也就是从当前研究活动中可以获取哪些信息。如图 8.1 所示，层级 4 涉及的内容通常是描述研究目的，介绍主要的研究活动或研究的主要发现，具体应写入哪些信息，取决于研究领域和期刊的要求。论文作者在层级 4 中呈现这些信息时，也有相当的灵活性，可以多关注自己领域内其他作者使用的语言，在平时阅读时有意识地积累相关句型，用在自己的写作中。请完成任务 8.9 和任务 8.10。

任务 8.9 层级 4 使用的句子模板

重新阅读表 8.1 中 McNeill et al.（1997）引言的层级 4，并翻开你选中的 PEA，阅读引言部分的层级 4，尝试提炼出几个句子模板，可以核对书后的参考答案。

任务 8.10 开始 OA 引言部分层级 4 的写作

请完成论文初稿（OA）引言部分层级 4 的写作，确保可以由层级 3 流畅过渡到这部分内容，为引言画上圆满的句号。读者在初见论文标题时会对文章的主要内容做出推测，建议在引言的层级 4 中使用论文标题涉及的所有关键词，满足读者的期待。

8.7 层级 5、层级 6：突出研究价值；预告文章框架

层级 5 和层级 6 均非引言的必备内容：书后论文范例中，只有两篇 PEA 的引言包含层级 5，强调研究价值；三篇 PEA 的引言均不含层级 6，没有明确说明文章的构思或结构。如果有必要在引言中使用层级 5，它可以出现在任何恰当的位置，层级编号只是用来命名，不代表顺序。但是 PEA 2 和 PEA 3 引言中的层级 5 都跟随在层级 4 之后，出现在引言的结尾，强调了研究活动或研究发现对各自领域的重要价值。

是否使用层级 6 取决于研究领域和期刊的要求，据我们观察，层级 6 经常出现在物理和计算机科学领域的论文内。在写作前，应多关注目标期刊内其他作者的做法。

8.8 引言部分的写作顺序

建议你采用下面的写作顺序完成引言部分。请务必在阅读前，确定自己已经与论文的合著者充分讨论，明确研究结果要包含哪些要点，以及这些研究发现对于目标读者的意义。

1. 先完成层级 4，即从研究目的写起，或是描述论文要完成或证明的内容。这可以说是引言中最容易写的一部分，通常出现在引言的最后一段，但是有必要在写作时优先完成。层级 4 应包含论文后面部分涉及的所有重要参数或问题，以便突出当前研究的价值和创新性，告知读者可以期待在后文中读到哪些内容。

2. 接下来完成层级 3，指出有必要做进一步研究，或存在研究空白。上面展示的例文中，层级 3 在引言多次出现，并且可以列出宽泛研

究需要下存在多个小的研究空白，然后从层级 3 过渡到层级 4。建议你在层级 3 的开头使用“however，although”等词；在指出研究需要时，使用“little information，few studies，unclear”或“needs further investigation”等标志词。

3. 然后思考如何切入层级 1，构建研究背景。应站在目标读者的角度进行考虑，寻找兴趣点或背景信息，结合论文标题要突出的重点，在开头就使用能迅速吸引读者注意力的词语或概念。

4. 随后，将已经从文献中收集到的内容编排到层级 2 中，也可以使用一系列“层级 2＋层级 3”的组合，用来说明层级 4 中不同研究活动的合理性。这部分内容至关重要，需要花费的时间也最长。同时需要进一步检索文献，看是否已从最大程度上掌握了当前研究领域内的相关发现，以及近期发表的相关研究。

5. 最后将所有层级整合成清晰连贯的引言。你或许需要添加一些语句，补充背景信息，还可能需要调整语句甚至段落的顺序，将逻辑理顺。写好引言的内容后，如果需要进一步修改，改善前后文的逻辑，请参考下面的 8.9 节。

8.9 文本润色，理顺写作逻辑

论文作者在使用英语写作时，应该对论证过程中逻辑的流畅性完全负责，确保读者可以正确理解。并非所有语言都要求如此。然而对于英语而言，母语作者想要达到上述目标也并非易事。下面，我们会为此提供一些重要策略。一些策略与前文有重复，但是在本节，我们整合了这些策略，方便你进行集中学习，并可以通过完成练习，巩固文本润色技巧。

策略 1：在开头就给出主题或观点

若想通过文字让读者在心中产生预期，需要文章使用含有充足信息的论文标题、小标题和引言。即：

A key to effective scientific and technical communication in English is to set up expectations in your reader's mind，and then meet these expectations as soon as possible.

如果目标期刊允许使用小标题，那就可以通过小标题不断向读者预告可以在后文读到哪些内容。同样，论文的标题就是从整体上，向读者介绍论文包含的关键信息。在段落层面，可以在段首使用主题句，告知读者当前段落的主要目的或观点。主题句也可以同时用来在上下文之间建立衔接。请完成任务 8.11。

任务 8.11 分析段落的主题句

阅读下列句子，这些句子都位于段首，你期待在随后的段落中获取什么信息？你认为相邻的上一段落是以什么信息收尾的呢？

1. Propagule pressure is widely recognized as an important factor that influences invasion success (MacDonald et al. 1989; Simberloff 1989; Williamson 1996; Lonsdale 1999; Cassey et al. 2005).

2. Two classes of putative Fe(II)-transport proteins (Irt/Zip and Dmt/Nramp) have been identified in plants (Belouchi et al. 1997; Curie et al. 2000; Eide et al. 1996; Thomine et al. 2000).

3. On Etna, the MAGFLOW Cellular Automata model has successfully been used to reproduce lava flow paths during the 2001, 2004 and 2006 effusive eruptions (Del Negro et al. 2008; Herault et al. 2009; Vicari et al. 2007).

查看书后三篇 PEA 的相应段落，看你的预测是否准确。同时可以核对书后的参考答案。

找出一篇你从未读过的论文，阅读引言所有段落的第一句话，你能否就此预测每段的内容？注意：段落的首句并不总是主题句，但大多数情况下如此。

策略 2: 从宽泛描述过渡到具体信息

英文读者总是希望在看到某个话题或观点的细节信息、具体事例等之前，先了解总体情况。

请阅读下面的段落，分析其内容是否符合“从宽泛到具体”的信息流动原则；同时观察是否有语句描述了宽泛信息，但出现在段落靠后的位置，如果有，请指出句子的编号。

① Pleuropneumonia (APP) can present as a dramatic clinical disease or as a chronic, production limiting disease in pig herds. ② A sudden increase in the number of sick and coughing pigs and a sharp rise in mortalities among grower/finisher pigs may herald an outbreak of APP in a herd. ③ On the other hand, signs may be limited to a drop in growth rate and an increase in grade two pleurisy lesions in slaughter pigs. ④ The disease surfaced in the Australian pig population during the first half of the 1980s and ten years later was regarded as one of the most costly and devastating diseases affecting the Australian pig industry.

比较过段落中的语句后，你是否同意句子④属于更具有概括性的内容，因而应优先出现在读者眼前？下面是本段的修改结果，为更好地满足信息流动原则，一些句子的措辞也稍作调整。

Pleuropneumonia (APP) surfaced in the Australian pig population during the first half of the 1980s and ten years later was regarded as one of the most costly and devastating diseases affecting the Australian pig industry. It can present as a dramatic clinical disease or as a chronic, production limiting disease in pig herds. A sudden increase in the number of sick and coughing pigs and a sharp rise in mortalities among grower/finisher pigs may herald an outbreak of APP in a herd. On the other hand, signs may be limited to a drop in growth rate and an increase in grade two pleurisy lesions in slaughter pigs.

策略 3: 将旧的信息置于新的信息之前

请通过下面的例子理解这条策略。对比下面两段话，虽然它们都包含了相同的信息，但采用了不同的语序，请判断哪一个版本更容易理解。

Version A ① Clay particles have surface areas which are many orders of magnitude greater than silt or sand sized particles. ② The ability of soils to shrink when dried is controlled by the interactions of these clay surfaces with water and exchangeable cations.

Version B ① Clay particles have surface areas which are many orders of magnitude greater than silt or sand sized particles. ② The interactions of these clay surfaces with water and exchangeable cations control the ability of soils to shrink when dried.

绝大部分读者会倾向于选择第二个版本（Version B)。原因如下：无论阅读哪个版本，当读到句子②时，意味着读者已经获得了句子①中包含的信息，此时，读者会将句子①视作旧的信息或已知信息。在这种情况下，第一个版本（Version A）的句子②以新信息开头，甚至要一直到后半部分才出现与已知信息构成衔接的词组“clay surfaces”，不符合新旧信息的编排策略，不利于读者理解。而第二个版本则在句子②的一开头就使用了已知信息，把新的信息调整到了句子的后部。请利用新旧信息的编排策略，完成任务 8.12。

任务 8.12 新旧信息的编排

下面段落中的哪句话有必要按照策略 3 进行调整?

Pleuropneumonia (APP) surfaced in the Australian pig population during the first half of the 1980s and ten years later was regarded as one of the most costly and devastating diseases affecting the Australian pig industry. It can present as a dramatic clinical disease or as a chronic, production limiting disease in pig herds. A sudden increase in the number of sick and coughing pigs and a sharp rise in mortalities among grower/finisher pigs may herald an outbreak of APP in a herd. On the other hand, signs may be limited to a drop in growth rate and an increase in grade two pleurisy lesions in slaughter pigs.

请核对书后的参考答案。

策略 4: 使用句首的 7 ~ 9 个单词在句与句之间建立联系

依然是上面的例子，另一个评价角度是：读者在句子②中何时才建立起与句子①的联系。在第一个版本中，读者需要读到第16个词，才首次看到熟悉的概念"clay"；在第二个版本中，则只需要读到第5个词，就能在两句话之间建立联系。为增强语言的可读性，我们建议在句子的前7～9个单词内就完成上述衔接，这样做有助于读者梳理文本信息。任务8.12中存在问题的句子，还可以进一步修改如下。

An outbreak of APP in a herd may be heralded by a sudden increase in the number of sick and coughing pigs and a sharp rise in mortalities among grower/finisher pigs.

修改后，第4个单词APP就是已知信息，确保了旧的信息位于新的信息之前。修改方法是将原句使用的主动语态"may herald"改为被动"may be heralded"。当段落内的逻辑需要调整时，这种方法非常有效。尽管有些写作手册会建议减少使用被动语态的次数，但在我们看来，确保信息流畅更为重要，因此必要时，应毫不犹豫地选用被动句。

策略 5: 句首的 7～9 个单词中应出现句子的主、谓语

请通过下面的例子理解这条策略。对比下面两句话，判断哪一句不容易理解。

① The definition of seed quality is very broad and encompasses different components for different people. ② The quality and quantity of flour protein, dough mixing requirements and tolerance, dough handling properties and loaf volume potential are quality parameters of wheat seed for bread bakers.

句子②不容易理解，读者要先读完长达19个词的主语，才能看到谓语动词"are"。这种句子的特点就是头重脚轻，主语太长导致句子启动太慢。下面列出两个修改版本。修改后，句子②的主、谓语都聚集在句首的7～9个单词内，同时新的信息——罗列出的具体"quality parameters"置于句子后部。

Edited version A ① The definition of seed quality is very broad and encompasses different components for different people. ② Quality parameters of wheat seed for bread bakers are the quality and quantity of flour protein, dough mixing requirements and tolerance, dough handling properties and loaf volume potential.

Edited version B ① The definition of seed quality is very broad and encompasses different components for different people. ② For bread bakers, quality parameters of wheat seed are the quality and quantity of flour protein, dough mixing requirements and tolerance, dough handling properties and loaf volume potential.

如果句子需要列出一系列信息，那么罗列的内容应置于句子后部，这可以视为一条通用规则。请完成任务 8.13 和任务 8.14。

任务 8.13 改写头重脚轻的句子

修改要求：句首的 7～9 个单词中应出现句子的主、谓语。

1. In this project the *Rhizoctonia* populations of two field soils in the Adelaide Plains region of South Australia were characterised.

2. A balance between deep and shallow rooting plants, heavy and light feeders, nitrogen fixers and consumers and an undisturbed phase is needed to achieve maximum benefit through rotation.

请核对书后的参考答案。

任务 8.14 理顺 OA 引言部分的逻辑

请参照本章介绍的策略，对论文初稿（OA）引言部分的文本进行润色。

9. 讨论部分

讨论部分将进一步向期刊编辑和审稿人展示论文研究活动的创新性和重要意义，在写作时，应侧重于当前研究发现与其他研究结果的对比，合理推测新研究发现的价值，同时说明当前研究的局限性。

9.1 确定写作结构时，应考虑的重要因素

在开始起草讨论部分时，应充分思考下列问题（本章主要针对的是独立存在的讨论部分，但如果其他小标题下的核心内容是对结果进行讨论，仍然可以参考本章列出的原则）。

讨论部分的结构

- 目标期刊是否允许把结果和讨论合并为一个整体部分（或自选小标题将二者的内容合并写作），并在其后设置独立的结论部分。这样的结构是否有助于你更好地讲述研究“故事”？
- 目标期刊是否允许在篇幅较长的讨论部分后面，增设独立的结论部分？这样做，对你是否有利？
- 目标期刊是否允许在讨论部分内使用小标题？这样的结构是否有助于读者在阅读时对研究结果的关键信息进行定位？

讨论部分与论文标题应紧密结合

- 尤其是确定出讨论部分要强调的关键信息后，应考虑是否需要对论文标题进行修订，使其更好地反映论文的重点。

讨论部分与引言部分应紧密结合

- 应确保讨论部分清晰回应引言中提出的问题，写作时，尤其要密切结合引言部分对“国家”的界定（8.2 节），指出研究空白（层级 3）时使用的论据，以及对研究目的或主要研究活动所做的描述。完成讨论部分的初稿后，应对照引言检验内容是否呼应。必要时，可修订引言部分的内容，确保讨论部分强调的关键信息在引言中也有所体现。
- 尽管如此，引言部分无需包括讨论结果时使用的所有文献，即：不必要的信息无需在论文的这两个部分中重复出现。

任务 9.1 讨论部分的结构

翻开你选中的 PEA，阅读讨论部分，回答下列问题。

- 论文是否包含独立的讨论部分？
- 讨论部分内是否使用了小标题？
- 讨论部分的内容是否出现在其他小标题之下？
- 论文是否包含独立的结论（或与之类似的）部分？

翻开 SA，做同样的练习。如果可能的话，与同事或老师讨论你的发现，并思考论文作者为何选取那样的写作结构？是否可以使用不同的结构，改进讨论部分？

9.2 为突出关键信息，应包含的内容要点

我们针对讨论部分应包含的内容要件，设计了下面的核查表。讨论某个结果时，不需要包含表中列出的全部内容，但建议你在写作时，花时间逐条思考下列要点，确保不会遗漏关键信息。

1. 重申主要研究目的、提出的假设或概述主要的研究活动。

2. 重申主要研究发现，通常按重要性排序，同时还应说明：

a. 研究结果是否能够验证研究假设，或研究结果是否能够回答研究问题、达到研究目标；

b. 研究结果与其他研究者的发现是否一致。

3. 基于其他相关文献，对研究发现进行解释；基于推测做出的论断，也需要有参考文献的支持。

4. 说明研究的局限性：研究结果能否在一定规模的领域中推广。

5. 说明研究的意义：归纳并推广研究结果，说明研究结果在更大的研究背景下具有怎样的价值。

6. 提出下一步的研究建议，还可以说明研究结果的实际应用。

上述 2～5 条的内容，通常在讨论几组不同的结果时，重复出现。

写作时，确定好你希望读者通过讨论部分领会哪些要点，并考虑使用小标题或主题句为读者定位出这些关键信息所在的位置。请完成任务 9.2～任务 9.4。

任务 9.2 讨论部分的内容要点

翻开你选中的 PEA，针对讨论部分，回答对应问题。

1. PEA1：阅读讨论部分第 2 个小标题 Specificity of GmDmt1;1 下面的内容，根据核查表，分析每句话所代表的内容要点。

2. PEA2：阅读讨论部分的第 1 段，根据核查表，分析每句话所代表的内容要点。

3. PEA3：阅读 5.3 Discussion 部分的第 1 段，根据核查表，分析每句话所代表的内容要点。

请核对书后的参考答案。

任务 9.3　分析 SA 的讨论部分

翻开 SA，选择讨论部分的 1～2 个段落，做如下练习。

- 根据核查表，分析每句话所代表的内容要件。
- SA 的论文作者是否使用了小标题或主题句等写作策略，定位讨论部分的关键信息?
- 关键信息与论文标题是否存在密切联系?

如果可以的话，与同事或老师讨论你的发现。

任务 9.4　开始 OA 讨论部分的写作

请开始论文初稿讨论部分的写作。合理利用 9.2 节中的核查表，确保自己论文的讨论部分没有遗漏相关内容。

9.3 准确传达观点的强烈程度

上一节核查表中的后四个内容要点都涉及对研究结果进行评述，应特别注意相关动词的使用。不同的动词形式可以传递出你对研究发现的不同态度以及观点的强烈程度。

如果句子包含“that”从句，作者可以选择在两个位置调整观点的强烈程度：

- 主句谓语动词和动词时态的选用；
- “that”从句谓语动词的时态。

表 9.1 中列出了一些选自 PEA 的例句。我们利用表格对句子进行了划分，并且用下划线标示了值得注意的关键位置。

表 9.1　在讨论部分的语句中选用动词、时态和情态动词

例句	主句主语	主句动词	“that”从句主语	“that”从句动词	句子其余部分
1	Our experimental results	demonstrate	that space- and propagule-limitation both	regulate	*S. muticum* recruitment.

续表

例句	主句主语	主句动词	"that"从句主语	"that"从句动词	句子其余部分
2	This	means	that running MAGFLOW on GPUs	provides	a simulation spanning several days of eruption in a few minutes.
3	These results	indicate	that *S. muticum* recruitment under natural field conditions	will be determined	by the interaction between disturbance and propagule input.
4	… it	appears	that GmDmt1;1	has	the capacity to function *in vivo* as either an uptake or an efflux mechanism in symbiosomes.
5	The presence of an IRE motif	suggests	that GmDmt1;1 mRNA	may be stabilized	by the binding of IRPs in soybean nodules when free iron levels are low.

例句1主句动词使用了一般现在时（属于永远正确的、强有力的描述），动词"demonstrate"本身传递出的意义也很强势，"that"从句中的谓语动词也是一般现在时。可以说，整句话都体现着论文作者对此观点确信无疑。这也说明如果作者认为论文中呈现的数据足够充分，完全可以在解释研究结果时使用最强有力的论述。

例句2与例句1的强度类似，"means"与"demonstrate"在信息的确定性上差别不大，主、从句也都使用了一般现在时。

例句3与前两句相比，强度下降。主句选用一般现在时，但动词"indicate"与"demonstrate"相比，确定性下降。"that"从句的一般将来时表示预测，且可能性很高。

例句4主句动词"appears"（通常跟随在形式主语"it"之后）的强度进一步下降。"that"从句使用一般现在时，体现的是前文出现过的论据本身的确定性。

例句5主句动词"suggests"还是强度不足，"that"从句由于使用了情态动词"may"，确定性也不高，因此例句5在表9.1中属于最不强烈的表述。但是，在讨论部分使用这样的句子不是坏事。在写作结果和讨论部分时，论文作者理应把观点、态度的强烈程度跟数据、论据的强度匹配起来，这可以通过在语句中调整动词和时态来实现。审稿人在评判稿件时，也会关注此类语言特征。接下来，请完成任务9.5。

任务9.5 利用动词调整观点的强烈程度

请通过调整划线动词，完成表9.2，并且将不同的动词形式按照观点的强烈程度排序，表格最下方给出的是最强有力的描述。

请核对书后的参考答案。

表 9.2 利用动词调整观点的强烈程度

The presence of an IRE motif	suggests ↓ demonstrates	that GmDmt1;1 mRNA	may be stabilized ↓ is stabilized	by the binding of IRPs in soybean nodules when free iron levels are low.	Weak ↓ Strong

在科技文体中，也存在不使用 that 从句的情况。请看出自 PEA 2 中的例句。

Previous studies have demonstrated a positive relationship between propagule pressure and the establishment success of non-native species.

该句谓语动词的宾语是一个名词短语（a positive relationship between propagule pressure and the establishment success of non-native species）。使用这样的句子结构，论文作者则不需要纠结 that 从句中到底该使用哪种时态。

请完成任务 9.6。

任务 9.6 分析讨论部分的观点强度，并付诸实践

翻开你选中的 PEA，找出讨论、结论部分使用了上述句型的语句，分析相关的动词，如果可能的话，与同事或老师讨论你的发现。

请认真思考自己的研究结果，并开始对研究结果进行评述，注意把观点的强烈程度与数据的强度匹配起来。

10. 论文的标题

你为论文选定的标题有着与读者交流的作用，这些读者除了期刊的编辑和审稿人，还包括文章发表后学术界的其他潜在读者。在第 3 章，通过分析审稿人评判稿件的标准，我们了解到论文标题应清晰传达文章的内容，但是做法不止一种。在本章，我们会为你提供一些建议，帮助你有效地吸引目标读者的注意力。

10.1 策略 1：简要地提供尽可能多的相关信息

论文标题需要抓住忙碌读者的注意力，让目标读者产生兴趣并获取整篇文章进行阅读。论文标题揭示出的信息越多，潜在的读者越能基于研究兴趣判断文章的相关度。许多期刊的投稿说明都有相关的规定，如"Journal of Ecology"需要"a concise and informative title (as short as possible)"；"New Phytologist"则要求"a concise and informative title (for research papers, ideally stating the key finding or framing a question)"，后面我们将进一步讨论论文标题的语法结构。

10.2 策略 2：突出关键信息

应确定出哪些词语（关键词）能吸引潜在读者的注意力，并把这些词置于论文标题的前部。这样做也有助于文献检索应用迅速搜索到你的论文，读者在使用文献检索服务时通常都会根据关键词筛选感兴趣的文章。建议你尽量把最重要的词语放在论文标题的起始位置，例如：

× Effects of added calcium on salinity tolerance of tomato

√ Calcium addition improves salinity tolerance of tomato

为了确保选中的关键词语能出现在标题前部，可以引入冒号（:）或破折号（—），在标点符号之前的主标题部分使用关键词，在标点符号之后使用副标题对主标题进行补充说明。下面的例子选自书后 PEA 的参考文献列表。

√ Disturbance, invasion, and reinvasion: managing the weed-shaped hole in disturbed ecosystems

√ Native weeds and exotic plants: relationships to disturbance in mixed-grass prairie

√ Methylamine/ammonium uptake systems in *Saccharomyces cerevisiae*: multiplicity and regulation

√ Resistance to infection with intra-cellular parasites—identification of a candidate gene

√ Mass flux measurements at active lava lakes：implications for magma recycling

10.3 策略3：使用名词短语、陈述句，还是问句

无论是论文标题还是文中的小标题，传统的做法都是使用名词短语：即中心名词加上一系列的限定、修饰成分。下面给出几个例子，加粗的单词就是中心词。

- **Diversity and invasibility** of southern Appalachian plant communities
- Food expenditure **patterns** in urban and rural Indonesia
- **Systems** of weed control in peanuts
- Iron **uptake** by symbiosomes from soybean root nodules
- **Evidence** of involvement of proteinaceous toxins from *Pyrenophora teres* in net blotch of barley

这些标题中有几个还是很好的：简明扼要，信息充足，并且把关键词放在了前部。但是，完全使用名词短语有时无法满足策略1和策略2的要求。上面列出的最后一个标题“Evidence of involvement of proteinaceous toxins from *Pyrenophora teres* in net blotch of barley”其实抛出了一个待回答的问题：到底是怎样的“involvement”？此外，标题开头的四个词都不够具体，无法吸引读者继续阅读。如果把这个标题改写为陈述句，效果可能更好。这其实也是审稿人给出的建议。（使用陈述句意味着包含主语和谓语动词，优势是可以给出关于研究结果的明确信息）

× Evidence of involvement of proteinaceous toxins from *Pyrenophora teres* in net blotch of barley

√ Proteinaceous metabolites from *Pyrenophora teres* contribute to symptom development of barley net blotch（Sarpeleh et al. 2007）

将论文标题写成陈述句适用于解决了某个具体问题并给出简单答案的情况。此时，我们可以避免在论文标题的开头使用“The effects of …”等模糊的短语。

× Effects of added calcium on salinity tolerance of tomato

√ Calcium addition improves salinity tolerance of tomato

如果问题的答案比较复杂，可以考虑把标题写成问句。

√ Which insect introductions succeed and which fail?

无论是使用名词短语、陈述句，还是问句，在做决定前都应该参考目

标期刊的常见做法或相关建议。根据我们的写作经验，在准备论文的草稿时，可以写出多个标题，在完成论文写作时，根据目标读者和文章的关键信息选出那个最合适的论文标题。

10.4 策略 4: 避免名词短语产生歧义

在标题中使用名词短语时，论文作者通常将大量信息包含在一系列的名词、形容词中，有时则会导致语义上的模糊——读者可以将标题做出多种解读。名词修饰名词最容易出现歧义，尤其是中心名词前面由名词短语修饰的时候。为了探究原因，我们首先来分析下面的例子。

名词短语“germination conditions”只可能有一种解读：“conditions for germination”，不存在歧义。类似的还有“application rate”，表示“rate of application”。但是“enzymatic activity suppression”的含义可能不止一种，到底是“suppression *of* enzymatic activity”还是“suppression *by* enzymatic activity”呢？为了避免歧义，我们建议你在使用名词短语时，把短语所包含的单词数量控制在三个以内，同时还要考虑该短语是否只表达一种含义。如果名词短语过长，要通过插入介词（of，by，for 等）对其进行改写，以澄清语义。例如：

× soybean seedling growth suppression

√ suppression of soybean seedling growth

注意：名词修饰名词时，用作修饰语的名词通常使用单数，请看下面的例子。

food for dogs → dog food

disturbance by herbivores → herbivore disturbance

nodules on soybean roots → soybean root nodules

请完成任务 10.1，巩固上面介绍的四个策略。

任务 10.1 分析论文的标题

完成表 10.1。如果可以的话，与同事或老师讨论你的发现。请核对书后的参考答案。

表 10.1 分析论文的标题

问题	PEA1	PEA2	PEA3	SA
论文标题是名词短语、陈述句，还是问句？ 论文标题的单词数量： 论文标题前部呈现的第一个概念： 你觉得论文作者为何将上述概念置于标题的开头？				

请思考能否进一步改进 OA 的标题。

11. 论文的摘要

11.1 摘要为何如此重要

- 忙碌的读者有时只会阅读论文的摘要（abstract 或 summary），除非摘要成功地吸引读者进一步完成全文的阅读。
- 资源有限的读者（所在的机构未订阅期刊数据库）恐怕只能获取到你论文的摘要。
- 文摘服务可能会选择将论文标题、摘要以及关键词录入其数据库，供用户检索。

11.2 关键词的选定

翻阅你的研究领域中其他相似的论文，查看这些论文使用了哪些论文标题范围以外的关键词。这些做法主要是可以帮助你了解相关的索引服务可能会使用哪些关键词。同时，有必要思考读者可能会对哪些词感兴趣，并使用这些词进行文献检索，试着从读者的角度做出预测。期刊编辑还会基于论文关键词选择审稿人，这是关键词的另一个重要作用，因此有必要在选定关键词时，考虑把这些词跟潜在审稿人的专业领域进行匹配。

11.3 摘要应包含的内容要点

一些期刊在指导作者完成论文摘要时，会提出一系列待回答的问题或给定一些小标题，有些期刊则不会这么做。但所有期刊都会限定摘要的字数（例如“The Plant Journal”的上限是 250 词，“Journal of Ecology”的上限是 350 词）。基于对科技领域大量论文摘要的分析，摘要应包含的内容要点概括如下：

一些背景信息	B
主要的研究活动（研究目的）及其范围	P
研究所使用的方法	M
研究得出的最重要的结果	R
陈述结论或给出建议	C

上述内容要件还可以进一步压缩，此时摘要的构成为：

主要的研究活动或目的以及使用的方法	P+M

研究结果 R
陈述结论（和建议） C

请完成任务 11.1。

任务 11.1 分析论文的摘要

翻开书后的三篇 PEA，阅读摘要部分，分析每句话对应的内容要点（即便你对某篇论文的领域并不完全熟悉，也不影响你完成上述任务。摘要通常很短，你完全能够对每句话做出判断，这对你自己的摘要写作也非常有帮助）。

请核对书后的参考答案。

注意：PEA2 出自期刊“Journal of Ecology”，该刊对论文摘要的要求如下：

Summary（called the Abstract on the web submission site）. **This must not exceed 350 words** and should list the main results and conclusions, using simple, factual, numbered statements. The final point of your Summary must be headed 'Synthesis', and must emphasize the key findings of the work and its general significance, indicating clearly how this study has advanced ecological understanding. This policy is intended to maximize the impact of your paper, by making it of as wide interest as possible. This final point should therefore explain the importance of your paper in a way that is accessible to non-specialists. We emphasize that the Journal is more likely to accept manuscripts that address important and topical questions and hypotheses, and deliver generic rather than specific messages.

上述要求的最后一句话对于分析 PEA2 很有指导意义。基于该刊对自己的定位，论文作者也就更清楚应该在论文的标题、摘要、引言的末尾和讨论等重要区域突出强调哪些内容。可见，目标期刊的投稿说明至关重要，你即将投出的稿件就是在回应投稿说明提出的要求。请完成任务 11.2 和任务 11.3。

任务 11.2 分析 SA 的摘要

翻开 SA，阅读其摘要部分，分析每句话对应的内容要点。如果可以的话，与同事或老师讨论你的发现。

任务 11.3 开始 OA 摘要部分的写作

请完成论文初稿（OA）摘要部分的写作。可以先根据内容要点，逐条进行写作，然后把写出的句子连接起来形成摘要的初稿。按照目标期刊的字数要求进行检查。如果需要对内容进行压缩，可以使用自己在分析论文摘要时观察到的写作手法。

12. 综述的撰写

我们关注的重点是科技论文写作，增加这一章内容使得本书的体系更加完整，同时也可以进一步比较综述和论文这两种类型的文章。

撰写综述类文章能为科研人员增加声望和被引用机会。由于综述的高引用率，在科研生涯早期就发表综述有利于研究者在事业上取得突破，甚至获得理想职位的面试机会。一些有影响力的重要期刊专注于发表此类文章，或设置了综述专栏。许多期刊发表综述时都会选择向作者约稿，但即便如此，正常的审稿环节也是必不可少的。如果没有受到邀请，也可以主动联系期刊的编辑部，询问目标期刊是否有意发表相关话题的综述文章。无论使用哪种方式，综述都需要涵盖全面、大量的文献资料，反映研究工作的最新进展，对所写专题进行归纳和评述，体现系统性与逻辑性。本章内容就围绕如何实现这些目标展开。

大多数科研人员都是通过文献综述才初次接近综述这一语类的，文献综述是硕士或博士阶段从事科研项目需要完成的内容，然后可能作为独立的章节出现在论文中。这样写成的文献综述，经过合理修改，有希望以综述的形式发表；如果计划周全，可以在刚开始写文献综述时就把未来的发表作为目标。发表出来的综述文章受众面更广，呈现文献资料的目标也有所变化（见注释 12.1），需要熟练应用科技文体的各种写作原则（见第 8 章）。当然，如果在论文写作过程中就充分考虑到上述因素将非常有助于改善论文的质量，尤其有利于论文的审核结果。研究者们曾采访过资深的论文评审人员，得到的结果是文献综述结尾的好坏往往决定了读者对论文的第一印象（Mullins & Kiley 2002）。因此，科研工作者都应该学习如何针对某个专题进行深入、有效的评述。

注释 12.1　综述写作：针对特定项目与针对学术期刊

为了更好地进行对比，一方面，我们需要了解高校、科研机构以及资助单位对于文献综述的要求，这类文献综述一般用于具体的博士科研项目或用于申报项目课题。另一方面，我们还需要考虑两类文章的不同读者群体以及评判标准。

	针对项目写作	针对期刊写作
读者	导师或学术委员会； 论文的评审人员； 资助单位	期刊的编辑和审稿人（从期刊读者的角度审查稿件）
评判标准	说服力和学术严谨性：论证是否有力，能否充分证明研究项目的必要性	创新性以及综述对该研究领域的价值； 作者的见解对研究工作是否有导向意义，需要指出未来的研究方向（通常不需要列举未来具体项目的细节）
相关的写作目的	逻辑线：To provide background information needed to understand your study; To show familiarity with the important research which has been done in your area; To establish your research as one link in a chain which is developing and expanding knowledge in your area. (Weissberg & Buker 1990) 逻辑线：In the review of literature "you review the primary literature on a particular topic, but you do so with a particular goal in mind: you wish to lead your reader to the inescapable conclusion that the question you propose to address follows logically from the research that has gone before." (Pechenik 1993) 研究空白：A literature review should serve to support, explain, and illuminate the logic behind the proposed research; it is used to explain the choices you have made in your research.	构建论点：The purpose of a good review is not to present a catalogue of names, dates and facts, but to present reasoned arguments about the field under review based on as many names, dates and facts as are necessary to support these arguments. (Lindsay 1995, 2011) 未来趋势：It is a critical analysis of relevant sources which shows what knowledge is still to be understood, or what research still needs to be done.

除了本书作者的观点，本章内容还融合了其他作者、综述文章审稿人以及杂志社的视角。我们曾经开展过“Literature reviews and review papers: Constructing compelling arguments”研讨会，帮助学员完成综述文章或修改其中的章节，在此过程中获得的经验我们也会一一呈现。本章选取的例子并非来自同一篇综述，而是选自一系列备受推崇的综述文章（综述类型：Tansley review），全部来自 New Phytologist Trust 网站（https://www.newphytologist.org/reviews）。你可以结合我们给出的文字片段体会综述的各种特征，然后选择其中一篇下载进行深入分析。从自己的研究领域或目标期刊中选择一篇综述进行剖析当然也是允许的。

12.1 编辑希望发表怎样的综述

"New Phytologist" 的主编 H. Slater 在与我们的访谈中表示，编辑们评判综述时，通常会退回那些没有新意的文章，尤其是跟已经发表过的综述相比，这些文章没有整合出新的内容或得到新的结论。除此之外，"New Phytologist" 在网站（http://www.newphytologist.com/authors）上还要求综述的作者 "Following a short introduction putting the area into context, and providing a 'way in' for the nonspecialist, these will concentrate on the most recent developments in the field"。因此，综述需要满足三个方面的特征：在引言中迎合读者，充分考虑读者的层次和需求；包含最近最新发表的文献；综合分析素材，提出新动向、新观点。我们从最后一个特征开始分析，那就需要再次提到关键信息（take-home message 或 THM）这个概念，THM 对于综述同样非常重要。

12.2 综述应传达的关键信息

在前面的章节中，我们曾多次提到论文的关键信息，论文作者可以将关键信息视为文章的思路框架，而所有被选中的研究数据则用来支撑这些信息或结论，为其提供证据。综述也是在向读者传递关键信息，但是"数据"的概念需要略作调整，我们可以把它理解为你带着批判性的眼光对文献资料所做的归纳和评论。作为综述的作者，你的任务就是合理地安排并呈现这些"数据"，传递前后连贯、条理清晰的信息。要确定出综述的 THM 并不简单，值得你深入思考。

可以通过如下几种思路形成你综述的 THM，这也间接决定了综述各个部分的写作顺序。

- 如果综述的作者一开始就有明确的论点，知道要引出的 THM 是什么（表 12.1 中的第一篇综述就是如此，我们曾采访过作者 John Harris 本人），那就可以把要点逐条列出，然后以这些要点作为综述初稿的提纲，下一步工作需要作者收集、组织论据来支持每个要点，最后形成一篇逻辑性较强的文章。
- 如果已经收集了许多证据，但 THM 暂时还无法界定，不妨先按照主题对素材进行分类。Lindsay（2011）指出，综述作者既要了解主题的细分，又要精炼观点，阐明结论。
- 如果已经明确了 THM，但尚未统一构思，可以自下至上进行推导。在一页纸的底部写下综述的 THM，然后逐层填写上级信息，每一层都要回答的问题是：说服读者相信这个结论需要哪些论据。另外，还需要花时间思考一系列小标题，贯穿文章的逻辑线。这些工作都完成后，综述各个段落的内容也就呼之欲出了。

形成自己的 THM 后，可以观察已经发表的综述，看 THM 在文中出现的位置。是一次性全部提出？还是出现在不同的位置，如何前后呼应，逐步论证，又是如何在文章结尾处加以强调和总结？THM 最可能出现在综述的标题、摘要、目录（小标题）、引言的结尾以及综述的结论部分。表 12.1 就选取了 Harris et al.（2011）和 Wymore et al.（2011）两篇综述的上述部分，请你认真对比。注释 12.2 和注释 12.3 是两篇文章的完整目录。

表 12.1 THM 在综述中出现的位置（请特别注意突出显示的文本）

具体位置	Harris et al.（2011）	Wymore et al.（2011）
综述标题	Modulation of plant growth by HD-Zip class I and II transcription factors in response to environmental stimuli	Genes to ecosystems: exploring the frontiers of ecology with one of the smallest biological units
摘要	Plant development is adapted to changing environmental conditions for optimizing growth. This developmental adaptation is influenced by signals from the environment, which act as stimuli and may include submergence and fluctuations in water status, light conditions, nutrient status, temperature and the concentrations of toxic compounds. The homeodomainleucine zipper (HD-Zip) I and HD-Zip II transcription factor networks regulate these plant growth adaptation responses through integration of developmental and environmental cues. Evidence is emerging that these transcription factors are integrated with phytohormone-regulated developmental networks, enabling environmental stimuli to influence the genetically preprogrammed developmental progression. Dependent on the prevailing conditions, adaptation of mature and nascent organs is controlled by HD-Zip I and HD-Zip II transcription factors through suppression or promotion of cell multiplication, differentiation and expansion to regulate targeted growth. In vitro assays have shown that, within family I or family II, homo-and/or heterodimerization between leucine zipper domains is a prerequisite for DNA binding. Further, both families bind similar 9-bp pseudopalindromic cis elements, CAATNATTG, under in vitro conditions. However, the mechanisms that regulate the transcriptional activity of HD-Zip I and HD-Zip II transcription factors in vivo are largely unknown. The in planta implications of these protein-protein associations and the similarities in *cis* element binding are not clear.	Genes and their expression levels in individual species can structure whole communities and affect ecosystem processes. Although much has been written about community and ecosystem phenotypes with a few model systems, such as poplar and goldenrod, here we explore the potential application of a community genetics approach with systems involving invasive species, climate change and pollution. We argue that community genetics can reveal patterns and processes that otherwise might remain undetected. To further facilitate the community genetics or genes-to-ecosystem concept, we propose four community genetics postulates that allow for the conclusion of a causal relationship between the gene and its effect on the ecosystem. Although most current studies do not satisfy these criteria completely, several come close and, in so doing, begin to provide a genetic-based understanding of communities and ecosystems, as well as a sound basis for conservation and management practices.

续表

具体位置	Harris et al.（2011）	Wymore et al.（2011）
引言的结尾	In this review，we present our account of the current knowledge that relates to the roles of the HD-Zip I and HD-Zip II TFs during plant adaptation under changing environmental conditions. We also briefly describe how these TFs integrate with phytohormonemediated responses，and indicate the limits of the current knowledge that relate to the mechanism of transcriptional activity and address the issue of how to overcome these limitations.	The major goal of this review is to explore how this concept applies to systems for which this approach has not been explicitly employed，yet，are sufficiently developed to explore broader basic and applied issues. We develop our ideas in the context of global change associated with commonly occurring，ecosystem-impacting events，including invasive species，climate and pollution. For example，in conifers，we explore how the interactions of foundation species（trees and squirrels）and climate can affect a much larger community. With examples from two highly invasive species that have become foundation species in their new environments，we explore how a single mutation in one example and a single haplotype in another example can have cascading effects to redefine their respective ecosystems. Similarly，with the release of endocrine-disrupting chemicals from human contraceptives into aquatic ecosystems，we explore how pollution can alter the gene expression of foundation species，which，in turn，may redefine these ecosystems. Thus，a community genetics perspective on interacting foundation species，exotics and pollution can broaden our understanding of how the genetics of foundation species can have unexpected consequences，and remind us of the complex connections that exist in both natural and exotic systems.
结论部分	Plants cope with a variety of environmental stresses by modifying their growth pattern to minimize the impacts of stress or to escape damage. The HD-Zip I and HD-Zip II TFs play an integral role in the signaling network that is triggered by endogenous and external stimuli，which leads to the modified growth characteristic of stressed plants（Fig. 3）. This growth adaptation is achieved through the regulation of cell differentiation，division and expansion by HD-Zip I and HD-Zip II TFs.	In summary，it seems that John Muir（1911）might have been correct when he stated that，‘When we try to pick out anything by itself，we find it hitched to everything else in the Universe.’ Numerous examples have emerged from diverse systems that make the case that the genes-to-ecosystem approach can provide an important perspective for the understanding of complex systems，for informing land managers and even for evaluating the effect on the human condition where the genetic impacts

续表

具体位置	Harris et al. (2011)	Wymore et al. (2011)
结论部分	However, there are still mechanistic and functional aspects of the HD-Zip I and HD-Zip II network that we need to understand before we are able to characterize the roles of each family in plant growth adaptation to environmental stresses. We know very little of the downstream genes that are ultimately regulated by the HD-Zip I and HD-Zip II TFs. Identification of their target genes is needed to build a comprehensive view of the regulatory pathways and will enable validation of the suite of promoters controlled by HD-Zip I and HD-Zip II proteins to define the cis-acting elements. More information is needed on the interaction profiles of HD-Zip I and HD-Zip II TFs, and the properties that specify interaction partners which contribute to cis element binding. The increasing amount of cell-specific microarray data available in public databases will also enable determination of potential interaction partners that are dependent upon expression in the same cell. If transcriptional regulation by the HD-Zip I and HD-Zip II TFs is conferred through a common cis element, that implies that there are factors determining the specificity of transcriptional activity, as transgenic plants over-or under-expressing HD-Zip I and HD-Zip II TFs have variant phenotypes. Elements contributing to the specificity of action remain elusive. Obtaining data on the three-dimensional structures of the HD-Zip I and HD-Zip II TFs will also decisively contribute to our understanding of the nature of the protein-protein and protein-DNA interactions.	of pollution can have unintended effects on the food supply and human health. Similar to Koch's postulates, we present four community genetics postulates for confirming or rejecting the hypothesis of a genetic effect on the community and ecosystem (Table 1): (1) the demonstration of a target species' impact on the community and ecosystem; (2) the demonstration of key traits that are heritable; (3) the demonstration of genotypic variation in the communities they support and ecosystem processes; and (4) the manipulation of target gene(s) or their expression to experimentally evaluate a community and ecosystem effect. The last of these is the least well documented (but see the exotic hydrilla example of Michel et al., 2004; Fig. 2). Nevertheless, as we genetically engineer and release organisms, the fourth postulate will be evaluated on a global scale. In complex systems involving many interacting species, we believe that there are three main advantages to this approach. First, the incorporation of a genetically based model places community and ecosystem ecology within an evolutionary framework subject to natural selection. Second, because a genes-to-ecosystem approach studies species within a community context, it is more realistic and less likely to result in management errors compared with a single species' approach. Third, the use of the genes-to-ecosystem concept can reveal important interspecific indirect genetic effects among species, thus generating meaningful applications for the conservation of biodiversity, restoration, bioengineering, climate change and even the understanding of important human diseases.

注释 12.2　Harris et al.（2011）综述目录

Modulation of plant growth by HD-Zip class Ⅰ and Ⅱ transcription factors in response to environmental stimuli

表 12.1 注释：如何有效地突出 THM

综述标题

两篇综述的标题都不是句子而是名词短语，因此可以向读者介绍话题的内容，但是不能陈述具体的结论。这可以说是目前最常见的标题形式，我们在近期发表的综述中（来自 New Phytologist Trust 网站），也找到一些采用提问方式写成的标题，如“Units of nature or processes across scales?

注释 12.3　Wymore et al. (2011) 综述目录

Genes to ecosystems: exploring the frontiers of ecology with one of the smallest biological units

The ecosystem concept at age 75"，以及"Innate immunity：had poplar made its BED?"。至于表格中的案例，Wymore 综述的标题使用了"frontiers of ecology"，突出话题的重要性；"exploring"相比"exploration of"而言，也更侧重动作。Harris 综述的标题指出了话题涉及的三项指标："modulation of plant growth；by (particular actors)；in response to (particular stimuli)"。就如第 10 章所述，标题是在宣传文章的内容，上面的两篇综述采用了不同的策略来吸引读者，一篇是在刺激求知欲望，一篇则对内容进行了细致勾画。为自己的综述制定宣传策略、选择标题是非常有趣的过程，我们的建议依然是一边写文章一边拟定多个标题，等到文章最终成型后，再选出最合适的那个。

摘要

我们在第 11 章介绍过论文的摘要，其内容要点包括背景信息、研究目的和范围、方法（方法论）、结果、结论（和建议）。综述摘要的内容要点可能需要有所调整，请完成任务 12.1。

根据我们给出的参考答案，表 12.1 中的两篇摘要都介绍了背景，其中一篇说明了目的和范围。虽然两篇摘要都未涉及方法部分，但其他的学科会有不同的要求，比如医学上的系统性回顾通常必须侧重研究方法。两篇摘要都有语句对应了结果部分，但我们将其界定为"information representing new synthesis or conclusions"。结论（和建议）也能找到对应的语句，一篇指出了领域中存在的问题，另一篇展望了未来的发展方向。由此可见，综述的摘要可能存在下列内容要件：

任务 12.1 分析综述的摘要

阅读表 12.1 的摘要部分，分析是否存在这些内容要点（Background；Purpose and scope；Method/methodology；Results；Conclusion/recommendation），并找出对应的语句。

请核对书后的参考答案。

- 背景（已知信息；已经达成的共识）；
- 目的和范围（非必备：可以说明本篇综述与前人的不同之处）；
- 方法（与综述的类型有关）；
- 概括主要内容，做出综合评价；
- 结论（存在的知识空白，新动向、新观点的重要性等）。

上面列出的内容要点可以用来分析你研究领域内的综述，同时记得查看自己的学科是如何在摘要中突出 THM 的。为综述撰写摘要时，也可以将上述内容要点作为核查表，指导自己的写作。

我们将在 12.3 节专门讨论综述的目录（大纲）。

引言的结尾

在第 8 章，我们为论文的引言划分了层级，其中的层级 4 非常重要，可能包含的内容有三类：描述研究目的，介绍主要的研究活动或研究的主要发现，但是具体需要写哪些信息，取决于研究领域和期刊的要求。我们还建议，完成 OA 时，有必要及早写出层级 4，甚至引言部分就要从层级 4 写起。以上所有内容都适用于综述的写作，请完成任务 12.2，分析两篇综述引言的层级 4。

任务 12.2 分析综述引言的结尾

阅读表 12.1 的引言结尾部分，回答下列问题。

1. 引言的结尾出现了层级 4 的哪些内容（可多选）？
   a. Statement of aim or purpose of the present work；
   b. Statement of main activity of the paper (what the paper does, or the authors do in the paper)；
   c. Summary of the findings or outcomes of the paper.
2. 综述中的哪些标志词可以帮助你做出判断？
3. 引言的结尾是否也出现了层级 6（图 8.1），说明了文章的构思，预告综述框架？你能否找到标志词？

请核对书后的参考答案。

表 12.1 用底色突出了有关 THM 的部分。有趣的是，任务 12.2 中第

二个问题的答案，也就是我们找出的标志词在表格中都没有突出显示，反而是跟在它们后面的名词短语和名词性从句代表了 THM 最重要的信息。通过分析我们还可以借鉴层级 6 的写法，让它脱离单调和无趣，并利用层级 6 细致勾画出综述的 THM。

结论部分

请回忆第 2 章给出的 AIBC（摘要、引言、正文、结论）结构，引言部分和结论部分的形状明显是对称的，说明二者需要相互呼应。表 12.1 刚好能让我们比较两篇文章的引言和结论。阅读表格相应的部分，或者拿出你自己准备的综述进行分析，回答下列问题。

- 引言结尾处提出的所有问题或要点是否都在结论部分以某种形式得到回应？
- 结论部分是否提出了某些新的问题或要点，没有在引言的结尾进行铺垫？
- 如果引言和结论存在不匹配的信息，你认为这是否会影响文章的逻辑性？这又是否会影响论证的力度？

我们在第 9 章为论文的讨论部分列出过相关的内容要点，而综述的结论部分可以借用其中的三项：

- 研究的局限性；
- 研究的意义（归纳出新的内容或得到新的结论）；
- 下一步的研究建议（和实际应用）。

Harris 综述的结论部分明确强调了后两项（意义：If … then …，以及未来研究方向）。Wymore 综述的结论则通过介绍优势间接地说明了研究的意义，同时也指出了局限性（The last of these is the least well documented. Nevertheless …）。

新手作者还需要注意一点：摘要和结论部分的措辞是不同的。我们仍然可以在表 12.1 中进行比较，尽管摘要和结论改变了许多说法，但二者确实是在对相同的概念进行讨论，这样写作可以保证两个部分的原创性，请你仔细分析和体会。为了使表达和用词更加多样，母语非英语的作者有必要收集几篇自己领域内高质量的综述文章，从我们讲解过的角度进行对比学习，准确掌握相关语句供写作时使用。

12.3 综述的结构

综述类文章使用的是 AIBC 结构（2.1 节）。总的来说，科技类的综述需要包括引言和结论，这两个部分与科技论文的对应部分极为相似（参考 12.2 节；表 12.1 注释）。正文部分按不同主题进行划分，根据具体的内容设置一系列标题或小标题（参考注释 12.2 和注释 12.3），方便读者阅读。精心设计过的小标题能够有效突出作者的论证思路。

有的时候，期刊会在摘要前面为读者展示综述的目录，那么你就需要了解允许呈现几级目录。例如“New Phytologist”只会呈现最高一级目录，因此 Harris 综述的目录（注释 12.2）不能完全出现在摘要之前，尽管他的二级标题包含了许多信息，但是读者在一开始是读不到的。相比之下，Wymore 综述的思路可以在一级目录（注释 12.3）中得以完整体现，读者也可以预先掌握文章的全貌。即使期刊不设置单独的目录部分，经验丰富的读者都会略读全文，优先查看文章中的标题、小标题，对内容和论证过程做到心中有数，所以无论如何，你都应该确保文章所使用的小标题可以形成有效的论证“路线图”。

正文部分的主题句：逻辑清晰的保证

根据 8.9 节策略 1：位于段首的主题句可以用来介绍观点或预告思路变化。对于综述而言，各部分之间有小标题提示论证“路线”，使用主题句则能让论证更加有力。为了更好地说明二者的配合，我们选取了 Wymore 综述正文的一个部分，注意观察小标题和段落首句的作用（见注释 12.4）。下面请回忆我们一起阅读 Wymore 综述的全过程：通过表 12.1，我们首先略读了综述中经常出现 THM 的几个部分，再具体到综述的正文，如果你想了解第五部分“Heritable traits，pine cones and climate”的主要内容，只需要阅读各段首句（注释 12.4）就能大致了解作者的论证过程。

注释 12. 4　综述正文中的小标题和主题句

V. Heritable traits, pine cones and climate

The level of serotiny, a heritable trait in lodgepole pine (*Pinus contorta*) stands, is influenced by climate, fire and seed predators, and, in turn, affects forest composition and dependent species' evolution (Fig. 3) . ...

Climate, through its effect on fire regimes, appears to exert a major selection pressure that acts on serotiny. ...

Seed predators also influence the level of serotiny in lodgepole pine stands (Fig. 3b) . ...

The interaction of fire, herbivory and serotiny cascades to affect the whole forest ecosystem through sapling density after a fire. ...

Geographic location and pre-fire levels of serotiny also explained much of the observed variation in biotic responses, including species' richness, abundance of opportunistic species, and cover and density of pine seedlings, forbs, graminoids and shrubs post-fire (Fig. 3d; Turner et al., 1997) . ...

The level of serotiny in a population also affects the evolution of individual species. ...

To summarize, climate affects fire regimes, which, together with seed predators, select for or against the heritable trait of serotiny in lodgepole pine stands. ...

The lodgepole pine system fulfills three of the four postulates.

好的主题句从来都无法一蹴而就，好的主题句也不是综述作者在完成初稿时顺手写出的，它们是通过文章的修改环节不断补充、精炼而成的。作者需要反复追问：自己想要表达的究竟是什么？然后把观点清晰地写出来。

12.4 综述中的表格、插图和注释框

综述中的视觉元素让作者能够清晰生动地分析、归纳文献素材，还能直观地呈现新机制、新过程的全貌。根据第 5 章的讲解，文章中的每个视觉元素都应该传递清晰的 THM，其文字说明也要精心撰写，强调同样的信息。请看我们选择的案例。

综述中的表格

表格能够综合大量数据并指向特定的观点，这其实要求我们设计出合理的表格，并仔细斟酌标题栏的内容。Kranner et al.（2010）在综述中使用的表格可以看出如下特征（http：//onlinelibrary. wiley. com/doi/10. 1111/j. 1469-8137. 2010. 03461. x/full)。

表题：Table 1　Examples of potential abiotic stress factors and their effects on whole plants and orthodox seeds，classified according to the eu-stress-distress concept

列标题：Stress factor；Effect on whole plants (Distress/Eustress)；Effect on orthodox seeds (Distress/Eustress)

行标题：Water deficit；Temperature；Fire；Nutrients；Wind；Contamination，for example by nonessential heavy metals

作者在正文中首次提到表格（共提到四次）时写道：

… As a result of their essential role in plant reproduction，one would intuitively expect that plants have evolved mechanisms that protect their seeds from stress. Indeed，in the dry，quiescent state，protected by their seed coat，many seeds are exceptionally tolerant of stress factors，such as temperature extremes，that are lethal to adult plants (Table 1) .

第二次提到时，作者写道：

In addition，severe stresses on the mother plant will generally cause distress for both orthodox and recalcitrant seeds (Table 1；Fig. 3) .

由此可见，作者在展开论证时不需要重复提及相关细节，而读者可以在表格的单元格中轻松找到相应的文献资料。

你所熟悉的 Wymore 综述也有效利用了表格，通过表 1，作者呈现了四种假设（Four community genetics postulates to establish a causal relationship between genes and their community and ecosystem consequences)，

这样做可以方便读者随时查阅这些内容。

综述中的插图

主要具有下列两种特征：

- 利用组合图形提供多项论据（引用文献、数据），支撑综述中的某一观点，例如 Wymore 综述的图 3 和图 4；
- 利用流程图或原理图说明新机制、新过程所包含的环节，例如 Harris 综述的图 3。

无论使用哪种图形，都应该确保图注说明的自明性，读者无需阅读正文的相关内容就能够理解，同时标明信息的来源。下面是 Harris 综述图 3 的图注说明：

Figure 3 The proposed plant developmental regulatory network involving homeodomain-leucine zipper (HD-Zip) Ⅰ and HD-Zip Ⅱ transcription factors (TFs). Black arrows indicate an interaction. The black dashed line represents a promoter, the cross-hatched box represents cis elements downstream of other trans-acting factors, the CAATNATTG box represents the HD-Zip Ⅰ and HD-Zip Ⅱ cis elements, the right-angled arrow represents the transcriptional start site, and the 'Downstream gene' box represents genes whose transcription will be either suppressed or activated.[1] The arrowed box represents post-translational modification that affects the ability of the HD-Zip TFs to dimerize and/or bind DNA (Himmelbach et al., 2002; Tron et al., 2002).

Harris 综述的正文在提到图形时写道：

… These observations suggest that a large network exists, where dimerization partners confer different transcriptional characters and two families compete over similar cis elements. This would enable environmental and endogenous signals to regulate fluxes in the network to establish developmental programs through differential gene expression (Fig. 3).

正文使用的动词“suggest”和“would enable”与图注说明的措辞(the proposed plant developmental regulatory network…）能够相互匹配，指出了这个调控网络处于何种现状；基于图注的描述，图 3 可以作为论据随时插入到必要的位置。

注释

当综述作者需要呈现与论证过程相关的话题时，可以借助注释来提供必要的背景信息。一部分读者是不需要了解注释内容的，而那些不熟悉当前特定领域的读者可能会对注释更感兴趣。注释通常以大量的文字为主，偶尔也会包含图、表等视觉元素，注释 12.1 就是一例。

12.5 综述内容核查表

我们把这一章介绍过的要点整理成了下列问题，在综述快要完成时，不妨核对这些注意事项，看是否有忽略的内容。

- 综述所涵盖的内容是最新的吗（如果你已经收到审稿意见，准备修改后再次投稿，那么投稿前务必要再次检索最新文献，了解学科进展）？
- 综述的标题能否准确反映文章内容，吸引潜在读者？
- 摘要是否包含了文章要强调的关键信息，说明了综述的价值？
- 综述的目录会单独刊出吗？是否选取了恰当的标题、小标题，向读者提示文章的思路？
- 引言部分是否考虑到了不熟悉当前专业话题或领域的读者？
- 综述的 THM 是否突出了新论点、新结论，加深人们对话题的理解？
- 综述的 THM 是否出现在恰当的位置并前后连贯，这些位置包括综述的标题、摘要、引言的结尾和结论，这四个部分所传递的信息是否一致？

12.6 综述的投稿与修改

综述的投稿

无论编辑是否约稿，投稿的注意事项都可以参考第 13 章的内容。编辑和审稿人的职责是筛选及时、全面的综述，为读者和学科带来崭新的视角和解读。在与编辑沟通时，要记得这一点：努力说服对方接收你的稿件，理由是你的综述提出了有价值的新内容。

撰写综述也是极为耗时的，对研究者而言还可能是额外的负担，因为研究人员本身通常就承担着科研项目以及其他职责。一旦决定要完成一篇综述，需要尽快动笔，免得其他作者抢先发表类似的文章，如果拖延太久，每隔一段时间又要重新对文献展开检索，实在是得不偿失。如果有编辑约稿，那更应该及时完成，因为编辑很可能会由于拖延太久而撤销邀请。

如果没有收到邀请，事先也没有询问过编辑部发表相关话题综述文章的意向，记得在投稿信中突出综述的价值（见 13.4 节）。在任何情况下，我们都建议你能按照投稿说明的要求准备稿件，并且参考我们在书中提供的策略，将录用的可能最大化。

回应编辑和审稿人

综述也是需要经过同行评议的。约稿甚至即使得到编辑单方面的正面答复也不能确保文章能够直接刊用或修改后刊用。

审稿人针对综述给出的反馈意见与针对论文给出的意见类型基本一致，只是就综述而言，“研究”和“数据”所代表的内涵有所差别。审稿意见的类型以及回应方式都可以参考 14.3 节以及表 14.2。

03

第三部分

论文发表

13. 投稿

向期刊投稿犹如参加一场竞赛，竞赛的裁判手中握有一套选拔标准决定你的输赢。为了提高获胜概率，你应该理解并尽量去满足期刊的选拔标准。准备稿件时，相关的标准（期刊的宗旨和范围、投稿说明等）可以在期刊的纸质版或网站上找到。还有一些标准关系到文章是否达到出版要求，这需要你熟悉期刊的编辑和审稿流程。接下来，我们从编辑和审稿人的角度为你介绍这些标准，帮助你更好地应对发表流程。

13.1 成功发表的五个秘诀

科研人员的成功很大程度上取决于研究成果的质量和数量，以及这些研究成果对其他研究者、研究活动的影响。科技论文是衡量科研人员水平的重要评价指标，而成功的发表者通常都采用了下列方法。

1. 帮助同事审稿，为期刊审稿，不断积累经验，在头脑中构建论文写作和审稿标准的框架。

2. 合理地规划研究活动和论文写作，按部就班地满足审稿人和期刊编辑对论文的质量要求。

3. 仔细选定目标期刊，按照其内容和格式要求准备稿件，将录用的可能最大化。

4. 投稿前，邀请同事使用结构化流程对论文进行评价，不断改进文章的质量。

5. 利用审稿人给出的建议修改论文，并向期刊编辑说明文章做出了哪些改进。

13.2 同行评议的过程

仅凭一篇科技论文并不能提出真理或验证某种必然，论文本身只是如实记录一系列观测活动、分析数据，并基于前人研究对结果做出解释说明。研究结果的可信度需要靠后续研究或应用不断验证，有些可以得到证实，有些会随着时间的推移而进一步修正。同行评议发生在文章发表之前，也就是更广泛的读者检验或使用某篇论文之前，这样做有助于保证科学界内研究的质量。可以说，同行评议是信息转化为知识过程中的一部分。论文作者、期刊编辑和审稿人共同协商文章的价值所在，三者之间的互动决定了论文提出的新信息是否值得推广和采纳，同行评议体系并不完美，但它确实为科学研究的出版标准做出了重要贡献。同行评议能够：

- 确保论文提出的假设得到合理的验证，同时确保论文得出的结果与研究所使用的材料、方法和分析工具相匹配；
- 确保论文作者基于研究结果恰当地表达观点、提出主张，客观地说明研究的意义；
- 结合期刊的标准，帮助其决定当前研究的焦点、创新性和重要性是否满足要求；
- 检查论文内容的呈现风格是否符合出版惯例，且便于读者阅读；
- 建议期刊编辑和论文作者在何处、如何对文章进行修改。

审稿人对期刊编辑来说至关重要，因为审稿人对稿件的质量评定起关键作用，而且大多数情况下审稿人都是免费为期刊提供专业意见。审稿人对论文作者来说同样重要，审稿人带着批判性视角检查论文的内容和写作，帮助作者更清晰地讲述研究故事，恰当呈现、突出要点。对于论文作者而言，同行评议是一个获取专业建议的好机会——领域内的专家会思考你的观点、理论、方法、结果、分析和解释，并做出点评。回应审稿人的意见其实也是在检验研究结果的可靠性，确立研究的效力，获得同行的认可。

理解同行评议的最佳方法就是加入审稿人的行列。如果你发表了自己的研究，那可能也有机会受邀成为期刊的审稿人。如果没有发表经历，你可以主动提出帮助同事审稿，或积极通过组会交流审阅已经发表的论文（请参考任务 13.2，更多审稿技能提升的策略详见第 16 章）。

一直以来，同行评议都由期刊主导，由它们邀请领域内的知名科学家担当审稿人，按照期刊的审稿要求完成相关工作。随着 Web 2.0 时代的到来，软件技术进一步发展，新的审稿模式也不断涌现，投稿前的同行评议出现更多可能，发表后同行评议也在开放的网络出版行业中流行开来。但同行评议本身依然在科技论文发表过程中扮演着重要的角色，只是其管理和操作流程或许会在未来有所改变。

13.3 编辑的职责

编辑人员需要通过出版符合期刊要求、具有科学价值的论文维护期刊的声誉。编辑在审稿人的协助下完成稿件的筛选，审稿人还可以提出改进建议帮助作者达到出版要求。期刊的编辑或相关人员在阅读来稿后，初步决定是否送交审稿人。编辑在遇到下列三种情况时，通常会直接拒稿：来稿不满足期刊的宗旨和范围，文章的语言或结构太差，或科学层面上存在明显缺陷（如何面对拒稿，详见表 14.1）。研究活动符合期刊的宗旨和范围、精心写作的稿件通常都会进入审稿流程；只是有些编辑可能会对来稿采取较高的筛选标准，不会轻易将稿件送交审稿人。你可以附上投稿信，向编辑说明自己的文章已具备审稿的条件。

13.4 投稿信

投稿信随着论文一起发出（或上传到期刊投稿页面的指定位置），你可以充分利用投稿信向编辑推荐自己的文章。在投稿信中，你将有机会感谢编辑的付出，并说明自己已经尽全力按照期刊的要求完成论文的写作。在投稿信中，你需要：

- 坚信稿件符合期刊的宗旨和范围，并表达自己的信心；
- 陈述论文标题和作者姓名；
- 告知编辑当前研究是新颖的，且不存在一稿多投的现象；
- 列举一些具体信息，突出研究的创新性和价值，注意不要单纯地复述论文的摘要；
- 对编辑可能提出质疑的内容做出解释，例如文章为何过长，或为何必须使用多张照片呈现重要的研究发现；
- 表示自己希望论文的结构和整体呈现能够令人满意；
- 表达自己期待收到审稿人的反馈意见。

图 13.1 是一封投稿信的范例，请在阅读之后，完成任务 13.1。

Date ...

The Managing Editor
Australian Journal of Botany
Address ...

Re Manuscript:
'Arbuscular mycorrhizal associations of the southern Simpson Desert'.
P. J. O' Connor, F. A. Smith and S. E. Smith

Dear Dr Zhu,

Please find attached the manuscript "Arbuscular mycorrhizal associations of the southern Simpson Desert". This manuscript examines the mycorrhizal status of plants growing on the different soils of the dune-swale systems of the Simpson Desert. There have been few studies of the ecology of the plants in this desert and little is known about how mycorrhizal associations are distributed amongst the desert plants of Australia. We report the arbuscular mycorrhizal status of 47 plant species for the first time. The manuscript has been prepared according to the journal's Instructions for Authors. We believe that this new work is within the scope of your journal and hope that you will consider this manuscript for publication in the *Australian Journal of Botany*.

We await your response and the comments of reviewers.

Yours sincerely,

P. J. O' Connor

图 13.1 论文作者写给编辑的投稿信

任务 13. 1 利用投稿信推荐自己的论文

阅读图 13. 1 中的投稿信，论文作者使用了哪些词组向编辑“强烈推销”自己的研究？把它们标示出来，然后核对书后的参考答案。

13. 5 审稿人的职责

同行评议时，编辑指定至少两位审稿人分别审阅稿件，评判来稿的质量、创新性、研究价值和内容的整体呈现。研究者通常是无偿承担审稿工作，提供专业意见，这些工作有助于推动相关的科学领域不断发展。审稿人往往：

- 对论文所属的整体研究领域是熟悉的（审稿人未必是论文特定主题上的专家）；
- 自己也在整体的研究领域内发表论文（可能是你某篇参考文献的作者）；
- 自己也忙于研究、写作、教学、管理和家庭生活；
- 愿意参与审稿工作，但时间和耐心有限；
- 对科学研究和写作有着自己的偏好甚至偏见。

Referee's Evaluation Form

General questions　　　　Reviewer number: ________

1. Is the contribution new ? ☐Yes ☐No
2. Is the contribution significant ? ☐Yes ☐No
3. Is it suitable for publication in the Journal ? ☐Yes ☐No
4. Is the organization acceptable ? ☐Yes ☐No
5. Do the methods and the treatment of results conform to acceptable scientific standards ? ☐Yes ☐No
6. Are all conclusions firmly based in the data presented ? ☐Yes ☐No
7. Is the length of the paper satisfactory ? ☐Yes ☐No
8. Are all illustrations required ? ☐Yes ☐No
9. Are all the figures and tables necessary ? ☐Yes ☐No
10. Are figure legends and table titles adequate ? ☐Yes ☐No
11. Do the title and abstract clearly indicate the content of the paper ? ☐Yes ☐No
12. Are the references up to date, complete, and the journal titles correctly abbreviated ? ☐Yes ☐No
13. Is the paper excellent, good, or poor ? ☐Excellent ☐Good ☐Poor

Please use a separate sheet for your comments.

Recommendation

☐ Accept without alteration
☐ Accept after minor revision
☐ Review again after major revision
☐ Reject

Reviewer's signature: ________　　Date of review: ________

图 13. 2 审稿人评判稿件时需要回答的问题

期刊可能会要求你在投稿时推荐潜在的审稿人，还可能是由编辑通过数据库或使用专业的网络选定审稿人。你不知道审稿人是谁，但大多数情况下，审稿人知道论文作者的姓名，当然这也取决于期刊的相关政策。审稿人需要阅读稿件，报告其质量，指出问题，并提出改进建议。审稿人还应填写论文质量评估表，确定文章是否值得发表，还是需要修改后重投。审稿人将报告书和评估表交给编辑，有时还会提交批注过的稿件（由于目前提交和审阅的都是电子版稿件，纸质版上的批注已经很少见了）。

期刊有各自的审稿要求，这些要求有时会发布在期刊官网上。如果你有同事担任过审稿人，你或许可以拿到那个期刊的审稿要求。下面是我们整理出的论文质量评估表，列出了审稿人在评判稿件时经常需要回答的问题（图 13. 2），图 13. 3 是一份审稿人的报告书，包含了审稿人的评审意见。请在阅读后，完成任务 13. 2。

To：Dr AB Brown，
Editor，*Journal of*...

Re：Manuscript Number...

Title...

Authors...

Dear Dr Brown，

The paper describes... . This is a topic which would benefit from additional work such as that described in the manuscript. However，a major concern with the paper is the interpretation and referencing of the literature in the Introduction and Discussion. Related to this is a lack of integration with previous work to explain aspects of the Methods. The paper needs re-interpretation after a thorough investigation of the literature. I recommend that the paper in its current form be rejected but believe that it may be suitable for your journal after major revision.

Introduction
The Introduction has incorrectly cited ［Brown et al. （1981）］ who actually showed that......

Methods
Factors relevant to the choice of Methods are：1）how old were the cultures that were used? 2）Does the age of the culture material affect the results ?

Results
The main claim by the authors that their Results showed that... is not correct. Their statement that the results show... needs correction.

Discussion
Relevant references seem to have been overlooked in both the Introduction and Discussion sections，including...

Other queries and suggestions are pencilled on the manuscript.

Yours sincerely，
CD Smith

图 13. 3　审稿人在报告中建议拒稿，但同时指出了大修重投后接收的可能性。报告删去了一些内容细节

任务 13.2 在组会中举办审稿活动

构建或加入由3～10位同事组成的科技论文讨论小组，并定期见面（例如每月一次）。选出一些最近发表的论文作为分析对象，每次组会讨论一篇。针对这篇文章，每位组员都可以利用图13.2中的问题汇总出一份报告书。在组会上，大家可以分析文章的优势和缺点，以及讨论可以如何对其进行改进（更多利用组会的建议，详见第16章）。

13.6 编辑如何决策

编辑收到审稿人的评审意见后，要决定如何答复论文作者。如果审稿意见不一致（尤其是只有两个审稿人的情况），编辑有时会将稿件送交第三个审稿人，寻求额外的建议。随后，编辑将所做的决定告知论文的通讯作者。回应编辑的信件也是一门技能，我们将在下一章详细讨论。

14. 如何回应编辑和审稿人

14.1 值得你参考的重要经验

人们总是很难接受、回应批判性意见，对于自己的研究或写作也不例外。在需要对编辑和审稿人的意见做出回应时，我们建议你参考下面的经验法则。

1. 下列两种情况都非常罕见：编辑/审稿人完全正确而论文作者完全错误；论文作者完全正确而编辑/审稿人完全错误。

2. 在回应审稿人时，既要对相关的意见做出妥协，又要保证不损害论文要讲述的研究故事或要传递的关键信息。

3. 要向编辑展示出自己会尽全力配合，改进自己的文章。

4. 退稿和受到批评不等于自己的研究没有价值或文章写得不够好。可以尝试向其他期刊投稿，在论文中加入更多的研究工作，或重写某个部分乃至整篇文章。

14.2 如何面对拒稿

一旦遭到拒稿，应找出具体的原因，以便决定下一步的工作。所有有经验的研究者都遭遇过拒稿，与这些资深的同事交流后，你将更好地理解——想要确立科学知识的效力，遭到拒绝是很自然的也是必要的。这些资深的同事或许还会告诉你，被拒稿件中的部分乃至全部数据最终都能得以发表。请记住：每个人都会经历拒稿。成功的作者之所以成功，在于面对拒稿以及录用时，都能够妥善地处理。表 14.1 列出了拒稿发生可能的原因和下一步的处理办法。

表 14.1 拒稿原因以及如何做出回应

拒稿原因	应对方案 1	应对方案 2
论文的内容不符合期刊的范围(例如文章内容可能过于专业、细化，聚焦在不恰当的主题领域，或不适合期刊的读者群体)	上述判断应该是编辑做出的(稿件还未送交审稿人)，因此有必要修改自己的稿件，并考虑投给更适合的期刊(参考表 1.1，查看备选期刊)	如果上述判断是审稿人做出的，结合 14.1 节中的经验 2 和经验 3，妥善处理审稿人在报告书中给出的意见，然后将修改后的稿件投给更适合的期刊(参考表 1.1，查看备选期刊)
论文在科学层面上存在明显缺陷	上述判断应该是编辑做出的(稿件还未送交审稿人)，因此要对稿件做出相应的修改(参考表 14.2)，争取能把文章最好的部分发表出来	
文章的语言或结构太差，无法送交审稿人	上述判断应该是编辑做出的，因此要对稿件做出相应的修改(参考表 14.2)，然后重新投稿或将稿件投向其他期刊	

续表

拒稿原因	应对方案 1	应对方案 2
高水平期刊本身的拒稿率就很高，即使审稿意见（整体上）是正面的	细读编辑的回复，分析其中是否有鼓励性的话语（如“revise and resubmit”）。如果编辑鼓励重新投稿，结合 14.1 节中的经验 2 和经验 3，妥善处理审稿人在报告书中给出的意见，然后将修改后的稿件重新投出	如果不鼓励重新投稿，结合 14.1 节中的经验 2 和经验 3，妥善处理审稿人在报告书中给出的意见，然后将修改后的稿件投向其他期刊（参考表 1.1，查看备选期刊）
审稿人没有完整阅读稿件或未能充分理解其内容； 审稿人提供给编辑的意见不够明确或遭到误读； 某些原因激怒了审稿人，毕竟审稿人是难以捉摸的，审稿人有时也会扮演不利的角色	你可以向编辑提出申诉，尽管成功率不高，除非审稿人或编辑在评判稿件时确实出现了重大失误。但无论是申诉、重投还是将稿件投向其他期刊，在此之前，都建议你结合审稿意见，合理地修改自己的稿件	修改、重投或将稿件投向其他期刊。应修改原稿中引起误解的内容，使其更加清晰明确。如果重新投稿到同一期刊，在投稿信中告知编辑曾经出现的误解，列出审稿人给过的正面意见，并说明稿件具体做出了哪些改进

14.3 如何应对“修改后刊用”或“修改后重投”

很少有稿件可以不经过修改就直接发表。需要修改的程度也各不相同，有时只需要改动语言、参考文献或格式，有时则需要大修后重投并再次由审稿人评议。期刊之间也竞争激烈，既要争取出版新鲜有趣的研究成果又要争取更大的订阅量，编辑们则期待接收高质量的稿件然后及时将它们发表在纸质版期刊中（或发布在网站上）。如果论文涉及的研究极具创新性，但稿件本身需要大修，编辑可能会拒稿并鼓励论文作者重写后再次投稿。如果稿件不需要进行重大结构调整、增加额外的研究内容或重写，编辑会选择“修改后刊用”，要求论文作者按照审稿意见修改稿件，并在指定时间内将文章返回。图 14.1 中展示的就是“修改后刊用”情况下，编辑的回信。这意味着接下来，你可以结合编辑和审稿人的意见对稿件进行修改。但问题是，理解或处理审稿意见也并不容易。

注意：“修改后刊用”或者“修改后重投”只能通过编辑回信时的措辞来判断。这也是你在收到编辑回复时的首要任务，只是判断起来也没那么容易。出于礼貌，或为了维持期刊在你心中的良好印象，编辑使用的语言可能非常婉转。如果难以做出判断，可以请同事一起来分析编辑的回复。不管怎样，在做出下一步决定之前，你都需要与论文的合著者商议如何做出回应。

回应审稿意见的方式不止一种，你也会形成属于自己的策略。有经验的作者通常会做出如下反应。

- 不要为这些审稿意见感到愤怒。审稿人或编辑可能确实对文章存在误解，也可能确实是你表述不清。应对审稿人的意见是论文发表的流程之一，别把它视为针对你个人信誉的攻击。

From: Dr AB Brown,
Editor, *Journal of*...

Dear Dr Zhu,

I enclose the referees' reports on your paper entitled... The referees agree that the paper contains much good material. However, they have recommended that it needs considerable revision before it can be published. In particular, I draw to your attention the following comments by the referees.

Referee 1:
- The Methods section does not give sufficient information, particularly about the sampling methods used.
- The results in Tables I and 2 are closely related and can be combined into a single table.
- The conclusion that there is a strong positive correlation between the number of organisms and soil salinity needs a stronger statistical basis.
- The results in Figure 3 are very preliminary-this really requires another survey. If this is not possible, the Figure should be deleted.

Referee 2:
- There are inadequacies in the Methods section, as indicated on the typescript.
- The Discussion is not well focused and does not include some important relevant publications, e. g. Jones et al. (2000). '......' in the *Journal of*...
- The conclusion is interesting but can be greatly strengthened. In particular, the findings are different from those of Walter et al. (1997) in the *Journal of*..., a study done in the USA. The work in your paper is in fact the first study of its kind outside Europe and North America and this should be highlighted.

There are other comments in the enclosed reports, and some corrections have been made to the English on the typescripts. If you can revise the paper along the lines suggested and resubmit by... then I will consider its acceptability for publication in the Journal without further reference to referees. However, additional refereeing may be necessary.

I look forward to hearing from you.

Yours sincerely,
AB Brown

图 14.1 “修改后刊用”的情况
(注意: 编辑的回复通常不会如此简短)

- 仔细阅读审稿意见并核对自己的论文，确保自己理解了审稿人或编辑的要求。
- 把难以回应或看不懂的某些意见标记出来。
- 可以与论文的合著者或其他同事进行交流，听取建议。如果你还是无法理解、不知如何应对，甚至感到愤怒，不妨暂时不去思考它们，过几天（不要超过一周）再着手解决，或许会有思路。
- 回顾 14.1 节列出的经验法则。
- 完成所有需要小修的项目（即未被要求重写的部分），然后记下每一处修改，在回信中向编辑说明。
- 回应主要的审稿意见时，可以参考表 14.2。

表 14.2　如何回应审稿意见、修改稿件

审稿意见	应对方案	可能需要修改的位置	对应本书的章节
研究目标不够明确	重写研究目标，把它们陈述清楚	引言部分（层级 4）	8.6 节
	确保研究目标与实验设计紧密联系	交叉核对引言部分（层级 4）与方法部分	第 7 章、第 8 章
	确保论文的讨论部分重新提及研究目标	交叉核对讨论部分与引言部分（层级 4）	8.6 节和 9.1 节
研究工作的理论前提或学术派别受到质疑	确保你已经对不同的理论做出说明（引用文献），并解释你的论文在验证其中某个理论	引言（层级 2 和层级 4；在讨论部分也可以再次强调）	第 8 章和 9.1 节
	如果论文在挑战已经获得公认的理论： ● 解释这条理论； ● 引用文献对该理论进一步探索； ● 借助特定的写作结构（小标题、主题句）展示自己的论证逻辑	引言（层级 2；在讨论部分也可以再次强调）	第 8 章
	如果有其他约束条件（研究结果不适用的情况），应做出说明	讨论部分	第 9 章
实验设计或分析方法受到质疑	阐述实验设计或分析方法的优势	在回信中为自己的文章进行辩护	14.3 节
	指出其他已发表论文中使用过类似的设计或方法（引用文献）	方法部分，以及在回信中做出说明	第 7 章和 14.3 节
	如果可能的话，补充关于实验设计或分析方法的额外信息	方法部分（在讨论部分也可以再次强调）	第 7 章、第 9 章
为改进文章质量，需要提供附加数据或其他信息	如果可能的话，提供附加数据	结果部分（也可能涉及其他部分）	第 5 章、第 6 章
	如果只要求进行小修，但又无法提供附加数据，在回信中向编辑说明情况	在回信中做出说明	14.3 节
	如果必须进行大修，但又无法提供附加数据，可考虑重写论文	论文的所有部分	第 4 章
需要删除部分信息，或讨论部分的某些内容	在不影响自己讲述研究故事的前提下删除相关内容。如果审稿人没有明确指出需要删减的信息，可以寻求同事的帮助和建议	论文的相关部分（通常是引言部分、讨论部分）	多个章节
	如果按要求删除内容后，影响到研究故事的完整性，重新评估编辑和审稿人对论文做出的正面评价，在回信中向编辑说明情况（强调正面评价），请求保留这些内容	在回信中做出说明	14.3 节
论文的结论不正确、说服力不足或过于绝对、强势	确保讨论部分与论文在一开始提出的研究目标紧密联系	交叉核对讨论部分与引言部分（层级 4）	第 9 章和 8.6 节
	重新评估你使用的参考文献（检查并引用相关的支撑文献），如果论据充足，在回信中向编辑说明情况	讨论部分，以及在回信中做出说明	第 9 章和 14.3 节
	确保所有的陈述都是合理的，观点的强烈程度也是恰当的	讨论部分	9.3 节
	如果有其他约束条件（研究结果不适用的情况），应做出说明	讨论部分	第 9 章

续表

审稿意见	应对方案	可能需要修改的位置	对应本书的章节
审稿人给出了其他负面评价，例如研究设计、文章写作或结构存在不足等	与同事共同讨论审稿意见	论文的相关部分	多个章节
	重写相关的内容，向编辑说明自己所做的每一处修改	在回信中做出说明	14.3 节
	说明自己已经尽力满足了所有的修改要求，在回信中使用一系列肯定句，例如“I have addressed point 1 by...”	在回信中做出说明	14.3 节
	如果论文的语言或语法受到批评，应寻求语言专家的帮助	论文的相关部分	第 17 章

审稿意见的主要类型

审稿意见不同，带来的挑战也不同。大多数审稿人的意见都可归入下面的七个类别。

1. 研究目标不够明确。
2. 研究工作的理论前提或学术派别受到质疑。
3. 实验设计或分析方法受到质疑。
4. 为改进文章质量，需要提供附加数据或其他信息。
5. 需要删除部分信息，或讨论部分的某些内容。
6. 论文的结论不正确、说服力不足或过于绝对、强势。
7. 审稿人给出了其他负面评价，例如“poorly designed, poorly written, badly organized, tables are too large, relevant literature not cited”或“English is poor”等。

针对某些不好理解的审稿意见，先判断它属于哪个类别。如果无法归入上面列出的情况，就需要考虑这条建议到底需要你做什么，然后看表 14.2 中列出的方案是否适用。如果编辑表示可以“修改后刊用”，那么审稿意见其实不影响论文最终的发表。在结合审稿意见修改文章时，要记得保持自己要讲述的研究内容的完整性，然后在此基础上采纳编辑和审稿人的建议。

表 14.2 列出了审稿意见的所有主要类型，并推荐了一系列应对方案。这些方案有的比较简单，有的则比较复杂，甚至需要从多个角度做出修正。然而，许多审稿人给出的意见都可以通过下面两种方式有效地回应，科技论文的作者都应该掌握它们。

- 引用已经发表的文献。已经发表的论文代表它们已经通过审阅并获得科学界的认可。因此其他作者的研究发现或结论，可以用作论据，帮助你通过对比来展开论证或形成论点。
- 改进文章的结构。本书已经详细介绍了科技论文各个部分的结构和写作逻辑。在修改论文时，可以复习相关章节，改进文章的结构、论证，应对审稿人的建议。

利用表14.2确定恰当的应对方案，并判断需要在论文的哪些部分做出修改（审稿意见也可能直接指出具体位置）。表14.2也根据具体的审稿意见指出了本书的对应章节，你可以重新阅读相关写作建议。请完成任务14.1。

任务 14.1 寻找真实案例进行分析

请有经验的同事分享收到过的审稿意见以及做出的回复。

1. 审稿人给出了哪种类型的审稿意见，结合表14.2做出判断。
2. 分析同事做出的回应是否与表14.2提供的应对方案一致。
3. 与同事讨论回复的思路和想法。

关于如何训练自己去满足审稿人的审稿意见，详见第16章。

将修改后的稿件返回编辑或重新投稿，同时附上回信

快速对审稿人和编辑的建议做出回应非常重要，无论你收到的决定是小修后接收还是大修后重投。与初次投稿时附上的投稿信类似，你可以为修改后的稿件附上回信，表达对编辑工作的理解和感激，并且说明已经尽力满足了编辑和审稿人的要求。在这封答复审稿意见的信中，你需要：

To：Dr AB Brown，
Editor，*Journal of*................

Re：Manuscript Number........
Title....................................
Authors................................

Dear Dr Brown，

Thank you for your letter accepting the manuscript entitled... pending revision. We have made all the changes you suggested in your letter and address all the comments of the two reviewers in the notes below. We have also attended to the formatting and language of the manuscript according to your suggestions. Please note that reviewer comments are shown in **bold** type and our responses in plain type.

We note that there was some disagreement between the reviewers about the usefulness of the section of the manuscript on'observer effects'and that only Reviewer #1 recommended that this section be dropped. We are concerned that omitting this section might contribute to a lack of transparency and repeatability. It is critical to deal with it，because without it our key result would be confounded. Also，in discussions with colleagues on this topic，observer effects are invariably a subject of keen interest，and we believe readers would be frustrated to have our approach to dealing with it relegated to a brief reference. We have made some minor changes to the'observer effects'section to shorten it. We would be willing to make further changes if you felt them necessary and would be grateful for your advice on the matter.

(cont.)

Response to comments by Referee ♯1

1. Survey site markers in Fig 2 are too small.

Survey site markers have been increased in size

2. How were $a_n(x)$ and $b_n(x)$ computed? If they were computed empirically this should be stated in the text.

Yes, $a_n(x)$ and $b_n(x)$ were computed empirically. The relevant section now reads: "The quantities $a_n(x)$ and $b_n(x)$ were derived empirically, by calculating, for each visit and both survey types, the proportion of patches in which x species had been seen by visit n. For example, after three different day surveys, there were eight patches in which 17 species had been discovered, so $a_3(17) = 8/38 = 0.21$."

3. The notation in the equations is very complex and as this paper may be of interest to practitioners it would be better to reduce the use of symbols in Equations (1)~(7).

The notation of Equations (1)~(7) comes from another paper, so must be left as is. However, we have eliminated the use of β in reference to statistical power, and just used the word 'power' instead.

Response to comments by Referee ♯ 2

All suggested corrections made by Referee ♯2 have been made in the text.

We believe the paper is now acceptable for publication and look forward to your response to the changes we have made.

Yours sincerely,
Dr Zhu

图 14.2 论文作者写给编辑，对审稿意见做出回应
（注意：案例经过改编，回复通常不会如此简短）

- 结合审稿人的报告书，分别列出你主要修改的内容；
- 说明自己也修改了小的错误（例如英语的语言错误）；
- 指出审稿人的正面评价以及不同审稿意见之间的矛盾（支持自己赞同的那位审稿人的立场，同时试着赢得编辑的支持）；
- 如果审稿人确实出现了错误，要为自己的文章进行辩护（或者借此机会引用其他研究者的重要论文，支持自己的观点）；
- 陈述你相信自己的研究非常重要，修改后的文章可被接收。

这封信最理想的结构就是把审稿人的意见复制一遍，然后逐条进行回应（可以把审稿人的意见加粗，以示区别），在回应中需要引述改写的内容或告知编辑修改的位置。记得复查修改后的稿件是否满足投稿说明的要求（如论文的格式、长度和风格）。图 14.2 中展示的就是为审稿人意见回信的案例。

将修改后的稿件返回给编辑，同时附上回信。

15. 论文写作流程

从完成论文的初稿到投出稿件涉及许多流程，不同的作者操作方法也各不相同。但是整体而言，论文的写作非常耗时耗力，即使对于已经完成的部分，也要做好返工的准备。论文的合著者会不断充实稿件，文章要讲述的研究内容也会更加完整、精彩，从初稿到定稿，多次修改是必不可少的。尽管如此，精简、高效的写作流程对每位作者都有利，我们建议你考虑下面的操作顺序。

15.1 前期准备与写作顺序

1. 筛选出一系列的研究结果，将其“打包”，用来驱动未来的整篇论文。收集相关数据，并与合著者讨论下列问题：

- 这些数据要说明的关键信息是什么（数据要讲述怎样的研究故事）?
- “打包”的数据是最值得关注的信息吗？是否需要为研究故事补充更多的数据？为了保证单个故事的连贯性，是否需要删减某些数据？
- 论文的目标读者是谁？数据所代表研究结果的价值如何？能否据此选定目标期刊？
- 写作任务如何分工（谁负责做什么）?
- 将谁列为作者，作者的姓名如何排序？需要向谁的帮助致谢？例如，国际医学杂志编辑委员会在网站上列举了论文作者的界定标准，请参考：http：//www.icmje.org/recommendations/browse/roles-and-responsibilities/defining-the-role-of-authors-and-contributors.html。
- 制定可行的时间表。合著者将在何时阅读稿件（确定好时间点后，可以把“回应合著者反馈”这一环节插入到下面的步骤中）?

2. 召集一些同事，简要介绍你的研究背景和理由（引言部分的层级2和层级3）、研究目标或假设、研究方法的框架、呈现研究结果所需的数据（图、表和文字描述），以及对结果的讨论和结果的价值。征集同事们的反馈意见，询问报告中不清晰的地方以及有关该研究的其他疑问。

3. 找出目标期刊的投稿说明，按照相关要求制作论文文件的模板。

4. 按照目标期刊的要求，编辑呈现数据需要的图和表。同时考虑所有图、表的必要性，是否需要合并或删减。确保每个图、表要传递给读者的关键信息都清晰可见，图表说明也包含明确的描述。在每一个图、表下方，逐条列出信息要点，在每一条要点旁，注明此内容可写进论文的哪个部分：结果（R）还是讨论（D）。

5. 完成结果部分的写作，突出论文的关键信息。

6. 仔细思考，逐条列出讨论部分要包含的观点。

7. 写出或修改论文的标题，确保标题反映出论文的关键信息。

8. 完成方法部分（或对应部分）的写作。

9. 完成引言部分的写作。按照层级 4-3-1-2 的顺序完成，如果需要层级 5，将其插入到引言恰当的位置，必要时，还可以在引言的最后增加层级 6（详见第 8 章）。

10. 完成讨论部分的写作。如果有单独的结论部分，也请在这一步完成。

11. 完成摘要的写作。

12. 选定论文的关键词。

13. 把完成的各个部分整合成论文初稿。

14. 按照 15.2 节的建议，对稿件进行复查和校对。

15.2 复查与校对

1. 不要立刻开始对初稿进行复查。研究表明，你至少要在 48 小时后才可以重读稿件，并复查你到底写了哪些内容，否则你读到的只是“你以为你写了哪些内容”。

2. 打印论文的纸质版进行复查，按顺序通读全文，判断是否需要对文章内容做出修改。遇到问题，不要停下来修改，在页面旁边留下标记或简要描述出现的问题即可。

3. 通读结束后，回到文章的开头，处理标记出来的问题。

4. 然后按同样的方法再复查一遍。

5. 按照上面的步骤，反复修改文章的内容，直到你对论文的科学层面感到满意为止。

6. 然后，修改文章的语言：调整逻辑的流畅性；语句之间、各部分之间的衔接。

- 在必要的地方，使用小标题提供有用的信息。
- 在段落中使用必要的主题句。
- 段落之间或语句之间是否从宽泛描述过渡到具体信息，并且将旧的信息置于新的信息之前（详见第 8 章）。

7. 检查拼写、标点和语法。

- 使用“查找”功能搜索自己经常拼错的词（例如“from”与“form”），并进行修正。
- 启用“拼写和语法”功能，但拼写检查是有局限性的。尽管单词拼写正确，但很可能用错了语境，例如将“their”写成了“there”，或将“its”写成了“it's”。拼写检查功能无法识别某些专业术语，需要你手动将它们“添加到词典”（添加前，应确保拼写无误），这样操作后，你就可以集中处理真正的拼写错误，也就是文档出现红色波浪线的位置。

- 检查标点和斜体字的使用，尤其是“*et al.*”和物种名称（是否需要使用斜体，请按照投稿说明的要求来操作）。
- 如果英语不是你的母语，在检查语法时要格外细心。可以采取这样的方式：准备直尺和论文的纸质版（不要在电脑屏幕上检查）。从论文每个部分的最后一句话读起，并且把直尺放在这句话下面，然后逐行向上，倒序阅读——检查每句话的语法，包括主谓一致、单复数形式、动词时态、冠词（a，an，the）的使用等。这样做是为了帮助你把注意力集中在句子结构上，而不是句子内容上。请注意，在此刻，你已经多次重复过上面的1～5步，因此，句子内容或者说论文的科学层面已经通过检验，你只需要解决语法问题。

8. 校对引用内容与参考文献是否对应，以及二者的准确性。

- 如果你使用了Endnote或Reference Manager等文献管理软件，这一步所包含的大部分工作都已经完成了。你还需要做的是检查软件输出的内容是否符合要求，如果输入软件的信息有误，那么输出时可能生成显示异常或不一致的条目。
- 如果未使用软件，在这一步需要认真检查下列几个方面：

ⅰ. 正文引用的文献都出现在参考文献列表中了吗？

ⅱ. 参考文献列表中的每个条目是否都在论文中出现过至少一次？

ⅲ. 文内引用和参考文献列表的著录格式是否符合期刊的要求（例如标点、间距、斜体或加粗、大写字母的使用等)？

9. 校对文章的排版：在“打印预览”模式下逐页检查，确保页面布局满足目标期刊的要求，各部分的小标题和文字内容一一对应，以及确认是否需要在每一页添加书眉标题。

10. 校对论文是否遵循投稿说明的格式要求，例如图表、图表说明是否应存放在单独的文件内，是否需要为期刊的网站提供附加数据等。

11. 完成上述工作后，再一次通读论文，排除可能遗漏的小错误。最好能请朋友或同事帮忙阅读与检查，你也可以在同事投稿前扮演同样的角色。

15.3 投稿前使用的核查表

接下来，你可以考虑寻求来自写作团队之外的反馈意见了。可以采取这样的方式：邀请有鉴赏力的读者阅读论文，并填写评估表对论文进行核查。表15.1是就是投稿前可以使用的核查表，也是基于本书讨论过的注意事项编制的，你可以访问本书的网站下载电子版（http：//www.writeresearch.com.au)，在结合学科特点和特定用途修改后使用。

表15.1 投稿前使用的核查表

	评估标准	反馈意见
1	论文标题是否准确反映了论文内容？	
2	论文标题是否把重要信息置于前部，吸引读者注意力？	

	评估标准	反馈意见
3	引言部分是否符合“从宽泛到具体”的信息流动原则，从宏观的话题切入然后过渡到论文关注的具体问题？	
4	引言部分是否在领域内最近的国际文献中有效定位出当前研究？	
5	引言部分是否明确指出当前研究要填补的知识空白(也就是需要说明为什么做这项研究)？	
6	在引言部分的结尾处，是否清晰陈述了研究目的或假设，概述了主要的研究活动或研究结果(具体应写入哪些信息，取决于研究领域和期刊的要求)？	
7	是否选用了恰当的研究方法(包括统计分析)来解决当前研究问题？	
8	材料和方法部分是否提供了足够信息来向读者证明研究结果的信度？	
9	结果部分是否回答了引言部分提出的研究问题，或达到了研究目标？	
10	研究结果的呈现是否有逻辑(是否与研究问题的提出、方法的呈现或讨论部分的顺序存在呼应)？	
11	为了讲述研究故事，所有的图和表都是必要的吗？能否合并或删减？	
12	图和表能否“独立存在”(读者不需要阅读论文正文中的描述就能理解图表信息)？	
13	独立存在的讨论部分是否在开头回顾了研究目标、假设或问题？	
14	研究结果是否与其他文献的相关发现进行了对比？你是否能提出其他的对比角度？针对比较的结果(相似、差异或其他)，论文是否给出了合理的解释或推测？	
15	论文作者是否对研究结果的价值、局限性以及对实际应用或未来研究方向的影响进行了恰当的说明？	
16	论文是否以结论部分或总结段落收尾，再次强调关键信息和当前研究的意义？	
17	参考文献列表是否完整(列表中的条目都在正文中引用过，并且正文中引用过的文献都出现在列表中)？	
18	文内引用和参考文献的著录格式是否符合目标期刊的要求？	
19	论文的摘要是否包含了目标期刊要求的所有信息，并且适度强调了主要研究结果及其意义？	
20	摘要是否遵循目标期刊的字数限制和格式要求？	
21	论文选定的关键词能否让潜在的读者迅速检索到这篇文章？	
22	对于改进这篇论文的质量，你还有什么其他建议？	

在使用过核查表以后，还可以邀请有经验的同事结合审稿标准为论文提供反馈意见，也就是模拟审稿人评判稿件的环节。你还可以考虑把图13.2中的评估表提供给这位同事。

经历过这轮反馈并最终复查过论文之后，就可以投稿了。祝你好运！

04 第四部分

写作与发表技能进阶攻略

16. 适用于个人和团队的技能提升策略

在研究小组、实验室或系内部可以采取多种策略、活动来训练写作与发表技能。一方面，资深的科学家可以担任组织者，鼓励或要求初学者参与进来；另一方面，学生以及刚开始从事科学研究的人员也可以一起组织有益的活动，并适时邀请有经验的作者分享经验。

如果英语不是你所在国家的工作语言，那么组织活动时是否使用英语可以根据实际情况来决定。在制定活动方案的阶段，可以听取英语老师的建议，一起讨论何时以及怎样提升英语能力，把相关的活动融入整体方案。还可以借助本书的体系，在活动时有计划地使用各章节涉及英语写作的内容，例如集体讨论某篇手稿时，或准备学术会议的幻灯片时。

接下来我们提供一些不同的活动方案，供你选用。无论采取哪种策略，都建议你可以在首次会面时，为每期活动设定时限（例如每两周举办一次，持续三个月，然后做一次回顾和总结），与组员共同确定活动目标，并且制定明确的基本规则。

16.1 组会交流

组会交流（journal clubs）可以有效增进组员们在某个具体领域内的知识储备，是科学界广泛采用的策略。通常，所有成员提前阅读同一篇论文（由组长指定或组员推荐），然后在见面时深入讨论。组员们可以轮流担任主持人，在讨论环节提出具体问题供大家交流。

每次组会结束前，还可以增加一个环节用来提升发表技能。结合本书相关章节在任务中提出的问题，组员们可以选择论文的一个或几个部分（标题、摘要、讨论部分等）进行分析，下面列举一些分析的角度：

- 论文的这个部分是否向目标读者有效传达了相关信息？
- 论文作者是如何做到的？
- 为了确保有效沟通，论文作者是否使用了本书介绍过的技能？你能否从中找出具体的例子？
- 论文还有哪些写作特色十分有效，让你印象深刻？
- 论文的这个部分存在哪些不足？
- 论文作者在文中是否强调了其研究活动与期刊契合的某些方面（结合期刊的出版宗旨和范围进行分析）？

16.2 建立写作小组

写作小组是一个宽泛的概念，组员们定期见面，共同努力在某个写作项目中取得进展。对于我们来说，这个写作项目更偏向于论文的草稿或某些章节。写作小组既可由资深作者带领或指导，也可以完全由新手作者自行组织。无论哪种形式，只要安排合理，选取恰当的工作、学习模式，都能带来益处。

由两至三人构成的写作小组应该是最简单的模式。组员们定期举行活动，阅读彼此的草稿，当面讨论稿件的优势和不足，并在讨论结束时约定下一步的写作、讨论计划。

16.3 尝试从不同角度提供反馈意见

在写作小组中，你可能会受邀点评他人的论文；在更正式的场合，你可能需要为学术会议或期刊担任审稿人。如果你没有接受过相关训练，可以通过下面的介绍，学习如何为他人的论文提供反馈。

在提供反馈意见之前，应明确对方需要你扮演哪种角色。有时候，一些“出格”的意见或许会被视为人身攻击，毕竟论文的作者在完成文章时曾投入大量精力。因此，如果有人要求你做出反馈，需要先问清对方的需求，以及希望你扮演的角色。

常见的需求是为论文的内容（科学层面）提供意见，这意味着对方不需要你点评论文的语言。尽管如此，这对于许多审稿人来说也并不容易。建议你参考投稿前使用的核查表（表 15.1），在表格的基础上做出反馈，不需要在稿件上进行批注。论文作者还可能在开始写作之前就寻求反馈意见，邀请你审核论文的核心内容（即文章的关键信息或要讲述的研究内容），此时，你可以借助任务 4.1 提出的论文起草过程中值得关注的问题，结合对方的回答，并在审核论文将要包含的全部图表之后，提供反馈意见。

如果对方要求你评审整篇文章，那么反馈意见也应包含多个方面。在这种情况下，你依然需要考虑自己的角色，可以从下列角度入手：

- 需要我提供多少指导意见（鼓励并建议对方做出改进）？
- 是否需要我扮演期刊的“过滤器”（决定文章的质量）？
- 是否需要我提供详细的指导（帮助对方掌握相关的技能）？
- 对方还有其他需要吗？

在思考过上述问题后，你可以进一步考虑自己提供反馈意见的出发点：

- 你要扮演专家的角色吗？能对疑问做出解答，并且要求对方采纳你的意见。
- 还是扮演一位相对资深的同事？能够基于自己的经验提供反馈，并要求对方重视你的意见。
- 还是以同伴的身份出现？在写作中同样面临各种问题，甚至也正在学习如何用英语进行写作，但可以胜任目标读者的角色，借助本书的知识与论文作者共同讨论改进的方向。

- 还是综合以上角色？在科学层面提供反馈时，偏向某种身份；从其他角度提供反馈时，尝试另外的身份。

出发点不同，反馈意见的措辞也会有差异。你能判断出下面四句话分别来自哪种角色吗？

- More explanation needed.
- Not sure what you mean here.
- Move this to the Introduction.
- This may fit better in the Introduction.

在提供反馈意见时，通常有如下反馈策略。请对照指出哪些描述比较符合自己的情况，又有哪些策略是自己愿意在将来尝试的。

- 先指出文章的优势，再说明哪些问题需要修改。
- 用不同的颜色标记不同类别的问题（例如内容方面的和语言方面的问题）。
- 仅指出最严重的问题。尤其是针对早期的草稿，在反馈意见中不会去纠正所有问题。
- 在反馈报告结尾处进行总结，指出论文做得好的地方和强烈建议改进的地方。
- 推荐其他资源（人脉、书籍、网站等）。
- 仅使用下列校对符号指出问题的类型，但不做修改（有时会附上修改后的版本）。

sp＝拼写（spelling）
p＝标点（punctuation）
sing/pl＝单复数形式（wrong choice of singular or plural form）
wo＝词序（word order）
agt＝主谓一致（agreement between subject and verb）
t＝时态（tense）
art＝冠词（a/an，the，or no article）
obn＝新旧信息排布（put old information before new information）

具体使用哪种反馈策略，也取决于多种因素：

- 你的资历（经验多少）；
- 你在工作单位或研究机构中的角色（职位要求）；
- 你的性格；
- 你与论文作者的关系；
- 论文作者的具体要求。

受邀为论文提供反馈意见时，在各种方式之间达到平衡很有挑战性，但是为此所付出的努力非常有价值，这有助于你从研究人员转变为资深作者，因为你掌握了一项重要的技能：为论文提供严谨、全面、有建设性的反馈意见。

16.4 成为审稿人

在16.3节中，我们讨论了青年科学家可以如何转变为学术会议或期刊的审稿人，然而在这种偏正式的场景下，许多问题也随之出现，尤其是审稿人将无法与论文作者直接沟通。审稿人的首要职责是协助期刊的编辑决定文章是否适合发表，在提供建设性意见时，也应侧重于提供改进建议，使文章满足发表要求。由于审稿人的反馈意见通常会原封不动地转交给论文作者，16.3节中介绍的原则和策略也就尤为重要了。

我们开展一系列的研讨会时，不会把为期刊或会议撰写反馈意见作为正式的训练内容；有些导师可能会在编辑允许的情况下，让其指导的研究生参与审稿报告的写作。更多人也开始意识到，为论文审稿并提供反馈意见需要经过系统化的训练。“British Medical Journal”就已经在其网站上提供了相关的培训材料（http：//www.bmj.com/about-bmj/resources-reviewers/training-materials），如果你想训练自己成为某个期刊或领域的审稿人，可以参考上述内容。

16.5 如何训练自己更好地回应审稿人

在前面的章节，我们已经介绍了一些回应审稿人的方法。除此之外，我们还推荐另一种训练方式，前提是研究团队中某位已经发表过论文的成员愿意分享发表过程中的全部文件。在我们看来，团队中这样的成员通常比较资深，且愿意帮助经验不足的后辈掌握相关技能。基于这些文件，可以开展如下训练活动。

1.论文作者将投稿时使用的原始论文和期刊的最初回应（编辑回信和审稿人报告）提供给全体参加培训的人员。

2.参与者们通读这些文件，然后以小组为单位，讨论准备如何回应编辑和审稿人的意见。每个小组还可以写下答复时要使用的句子，并记录对原始论文所做的修改。

3.每个小组向全体人员汇报各自的讨论结果。接下来，论文作者向大家说明自己当时的做法，并将真实的回信提供给所有人。如果可能，还可以向大家补充描述论文的通讯作者以及写作团队其他成员看到审稿意见时的情绪反应，以及大家当时是如何处理这些情绪的。

4.大家再一次分组讨论，阅读回信，对比作者的实际回复与上一轮小组讨论结果的异同，并分析差异出现的原因。

5.随后，论文作者为大家点评每个小组提出的问题。

6.如果论文经历了第二轮评审，可以根据实际需要，重复上面的环节，看能否为参与者们提供新的领悟或更有价值的视角。否则，只需要邀请论文作者解释论文经历的后续环节和最后的录用结果即可。

7. 培训结束后，参与者要从以下两个方面总结经验：如何使用恰当策略回应审稿意见；以及在投稿前，应如何复查校对，需要特别注意哪些问题。

8. 如果参与者的母语不是英语，还可以通过培训，收集步骤 4 和步骤 6 中涉及的有用的句式或表达，在自己答复编辑和审稿人时使用。

17. 提升专业英语技能

17.1 何为专业英语

可以说，你用来描述自己研究活动所使用的英语是某种具有专门用途的英语，例如海洋生物学英语、植物生物技术学英语等。这意味着即使英语是你的母语，在刚开始接触到某个研究领域时，都需要学习并掌握那个学科的专业英语。在前面的章节中，我们从不同的方面介绍了科研人员在写作论文的各个部分时应如何使用英语。在这一章，母语非英语的作者是我们的关注对象，这些科研人员可以从哪些角度改进英语的语法和语言的使用呢？我们将从论文中常见的错误类型开始分析，不同的错误会从不同程度上影响编辑和审稿人对文章的看法。随后，我们将为你介绍提升专业英语写作技能的实用策略：句子模板提炼和语料库检索软件 AdTAT（Adelaide Text Analysis Tool）的使用。接下来，我们希望能帮助你攻克科技写作中的语法难题，尤其是母语非英语的作者容易感到困惑的地方。希望以上三项内容可以满足你的实际需求。

17.2 常见错误类型及其严重程度

科技写作最需要清晰的表达，因此有必要思考哪类错误对语义的影响最为严重，请完成任务 17.1。

任务 17.1 常见的错误类型

1. 与他人讨论（或自己整理）：如果你是某国际期刊的编辑，在收到母语非英语的研究人员投稿时，你认为稿件中可能出现哪些语言问题？

2. 我们列出了常见的一些错误，请将这些错误填入表 17.1 中，按照每种错误对语义表达的影响程度分类：基本不影响；影响；严重影响。

可供选择的错误类型：

① 单复数形式有误（例如：all tea leaves sample were oven dried）

② 句式过于复杂或不够准确（例如：This may be due to lower pH hinders dissolution of soil organic matter and decreases total dissolved Cu concentration because of Cu-organic complex reducing.）

③ 主谓不一致（例如：the results of this study suggests that...）

④ 介词使用有误（例如：similar with the results of other researchers）

⑤ 冠词 a/an/the 使用不当（例如：the accumulation of Cu in human body）

⑥ 情态动词使用有误（例如：would 和 will 的误用，can，could 和 may 的误用）

⑦ 词性有误（例如：drought resistance varieties）

⑧ 时态选用不合理（例如：描述已经得出的实验结果时使用现在时态）

表 17.1 常见错误类型及其严重程度

Rarely/slightly affects meaning	Sometimes/moderately affects meaning	Often/seriously affects meaning

分类结果以及论文读者对这些错误的看法，请查看书后的参考答案。

3. 请将所有的错误类型重新分类，按照你准备在写作中避免的优先级别排序：高度优先；中等优先；较后考虑。针对每种错误，可以采取怎样的改正策略，请查看书后的参考答案。

期刊的编辑怎么说？

期刊的编辑们非常关注稿件的沟通性。尽管有些期刊会聘用文字编辑，但论文的作者才最应该为文章的语言质量负责，尤其是论文使用的英语要清晰易懂。对于编辑来说，论文的学术价值（科学层面）最为重要，但所有科学内容都需要清晰易懂的表达。下面几段话引自 Elsevier 在线编辑论坛（www.elsevier.com/wps/find/editors.editors/editors-update/issue10d，2008-1-16）：

This is a long-standing problem. In the past it was solved to a large extent by detailed copy-editing of accepted papers. I became aware that this was apparently no longer being done when papers started appearing with ungrammatical titles.

For the researcher and for the reviewer, we should emphasize the scientific contents of their work. Language skills should not be the barrier.

The Authors may have important data, which is useful for the Community, and must be helped.

由此可见，

• 论文的学术价值最为重要；

• 但论文的内容必须清晰、易懂。

许多期刊在其官方网站上都明确指出：论文的作者需要对文章的语言质量负责。同时，出版社也建议：投稿前，母语非英语的作者应该使用在线编辑服务或出版社推荐的付费润色网站改进论文的语言。但即使你准备使用类似的服务，在此之前，你仍然需要尽全力完善论文的语言，使其符合英语的一般规则，并且表述清晰、无歧义。这样做不仅能为你节省开支（文字编辑一般按服务的时间收费），还能确保论文交由编辑处理时，不会因为修改语言而改变了文章的本意。

针对上面的问题，我们提供两个建议。

• 在撰写论文时，优先使用短句（若使用从句，数量不要超过两个）。如果将来有需要，再把短句连接起来。

- 根据特定的学科，积累适用于不同场景的表达，构建自己的资源库。

下面几节就介绍了构建资源库的方式。

17.3 合理的语言再利用：句子模板

研究表明，母语非英语的作者在发表论文时，常常会重复使用相同研究领域其他论文中的语言。这种策略何时是合理的，何时可能被视作“文本抄袭”，许多研究者都对此进行了讨论（Flowerdew & Li 2007）。然而有一点可以确定，论文内容（科学发现）和语言（传达内容的载体）在科学写作中的地位并不平等。科研工作的原创性在于内容，在于数据以及对数据的分析、阐释，而不在于语言。这与人文、社会科学领域中的观念有所不同，这些学科通过语言来形成论点，因此语言本身就是论文的内容。尽管如此，避免有剽窃的嫌疑依然是用英语进行学术写作的基本原则。在引用他人的研究发现或结论时，应使用自己的话改写原文，并标明信息来源。下面介绍的方法能帮助你合理利用其他论文中的语言，你甚至可以把一些好的语句收集起来，提炼成句子模板，供后续写作使用。模板的制作要领在于删掉“内容区段”（即名词短语），保留原句的框架。

为了理解“句子模板”的概念，请先阅读下面的段落。该段落用于阐明研究目的，节选自 Li et al.（2000）的论文“Water use patterns and agronomic performance for some cropping systems with and without fallow crops in a semi-arid environment of northwest China”。

As part of a long-term research effort aimed at establishing a sustainable rainfed farming system in the semi-arid and sub-humid regions of northwest China, this paper presents a detailed study on the water use patterns and agronomic performance for some cropping systems with and without fallow crops in a semi-arid environment. The objectives of this study were to: (1) determine the grain and aboveground biomass production and water-use efficiency of individual crops grown in the rotation; (2) analyze the seasonal and inter-annual patterns of soil water storage and utilization as well as water stress for the four major rotation crops such as winter wheat, corn, potato and millet; (3) determine the grain and aboveground biomass production and water-use efficiency for different rotation systems and evaluate the capacities of the rotation systems with and without fallow crops to utilize soil water storage in conjunction with seasonal precipitation; (4) establish whether the introduction of fallow crops into the wheat monoculture significantly influences the quantity of water stored in the soil that will be used by the subsequent wheat crop; and (5) discuss the characteristics of soil conservation for different rotation systems.

若删去段落中与当前研究相关的所有名词短语，可以得到一系列框架，这些框架就是我们所说的句子模板。

As part of a long-term research effort aimed at ~~establishing a sustainable rainfed farming system in the semi-arid and sub-humid regions of northwest China~~, this paper presents a detailed study on ~~the water use patterns and agronomic performance for some cropping systems with and without fallow crops in a semi-arid environment~~. The objectives of this study were to: (1) determine ~~the grain and aboveground biomass production and water-use efficiency of individual crops grown in the rotation~~; (2) analyze ~~the seasonal and inter-annual patterns of soil water storage and utilization as well as water stress for the four major rotation crops such as winter wheat, corn, potato and millet~~; (3) determine ~~the grain and aboveground biomass production and water-use efficiency for different rotation systems~~ and evaluate ~~the capacities of the rotation systems with and without fallow crops to utilize soil water storage in conjunction with seasonal precipitation;~~ (4) establish whether ~~the introduction of fallow crops into the wheat monoculture~~ significantly influences ~~the quantity of water stored in the soil that will be used by the subsequent wheat crop~~; and (5) discuss ~~the characteristics of soil conservation for different rotation systems~~.

保留下来的模板可以写成（NP 即名词短语）：

As part of a long-term research effort aimed at [NP1], this paper presents [NP2] . The objectives of this study were to: (1) determine [NP3]; (2) analyze [NP4]; (3) determine [NP5] and evaluate [NP6]; (4) establish whether [NP7] significantly influences [NP8]; and (5) discuss [NP9] .

注意：当这些框架与你要表达的意思契合时，再套用句子模板。为更好地记录句子模板所表达的含义，可以单独列出那些删掉的名词短语，并描述其特征，如表 17.2 所示。

表 17.2　可填入句子模板中的名词短语及其特征

名词短语	特征
1. establishing a sustainable rainfed farming system in the semi-arid and sub-humid regions of northwest China	verb+ing + NP + in + [NP of location]
2. a detailed study on the water use patterns and agronomic performance for some cropping systems with and without fallow crops in a semi-arid environment	*a study* + on + NP + in + [NP of location]
3. the grain and aboveground biomass production and water-use efficiency of individual crops grown in the rotation	NP + of + [NP referring to features of study already introduced]
4. the seasonal and inter-annual patterns of soil water storage and utilization as well as water stress for the four major rotation crops such as winter wheat, corn, potato and millet	NP + for + NP stating subjects of study
5. the grain and aboveground biomass production and water-use efficiency for different rotation systems	NP + for + NP stating subjects of study
6. the capacities of the rotation systems with and without fallow crops to utilize soil water storage in conjunction with seasonal precipitation	*the capacities of* [NP] to + verb + object

续表

名词短语	特征
7. the introduction of fallow crops into the wheat monoculture	*the introduction of* + NP + into + NP
8. the quantity of water stored in the soil that will be used by the subsequent wheat crop	NP of measurement
9. the characteristics of soil conservation for different rotation systems	NP referring to types of conclusions expected from the study

建议你在阅读论文时寻找有用的句子模板，不断充实模板库。尤其是当你把论文的内容读完后，额外用10分钟左右的时间来确立值得再次利用的句子模板，可以按照它们所在的论文各个部分进行分类，然后整理到专门的文件或记录本中。请完成任务17.2，这个任务是围绕引言部分设计的。

任务 17.2 为引言的层级 4 制作句子模板

1. 查看书后三篇PEA的引言部分，找出层级4的相关语句。提示：作者在层级4中使用非常具体的语句陈述研究目的，或说明主要研究活动或发现。这些语句使用了怎样的框架？写下你提炼出的句子模板，并与书后答案进行对比。

2. 分析自选论文（SA）引言中的层级4，若有合适的语句，提炼句子模板。列出在说明研究目的时涉及的名词短语，记录它们的特征，这些信息可以帮助你在将来更准确地套用句子模板。

17.4 名词短语再揭秘

描述一项研究要用到大量的名词短语，这些名词短语都是专业特有，且至关重要。想要提高专业英语，你需要专门花时间学习它们，并精准记忆。我们将在这一节，剖析名词短语的结构，同时向你解释“名词修饰名词”这一难点。

名词短语中不包含谓语动词，所有的部分都围绕一个中心名词，对其进行修饰、限定。下面给出几个例子，斜体字就是中心词。

- the *mechanisms* of salt marsh succession
- *interactions* involving carbohydrates
- the seasonal and inter-annual *patterns* of soil water storage and utilization

请注意，较长的名词短语可能由几个名词短语构成，中间用介词连接。

一个特例：名词修饰名词

我们发现：这种结构对于母语非英语的作者来说是一个难点。resource availability 就是使用名词修饰名词，解释成 availability of resources，但是在科技文体中，将 resources 提前，把名词短语简化，是常见的做法。此时，介词得以省略，用作修饰成分的名词改为单数（通常也不使用所有格）。再如，carbohydrate interactions 表示 interactions involving carbohydrates。我们从三篇 PEA 中提取了几个名词短语，请阅读表 17.3。

表 17.3 PEA 中的名词短语（名词修饰名词）

名词修饰名词	扩展形式
propagule pressure	pressure exerted by propagules
invasion success	success of invasions
field work	work conducted in the field
urchin disturbances	disturbances caused by urchins
legume root nodules	nodules on the roots of legumes
bacteroid activity	activity by bacteroids
bacteroid iron acquisition	acquisition of iron by bacteroids
soybean homologue	homologue in soybeans
lava flow hazard	hazard from flows of lava
a hazard map	a map of hazards
discharge rate estimates	estimates of rates of discharge

记忆这种结构时，可以参考 food for *dogs* is *dog* food，请按照这种思路完成任务 17.3。

任务 17.3 还原“名词修饰名词”

你的研究领域中有哪些常用的名词短语？举出三个“名词修饰名词”的例子，并使用扩展形式解释每个名词短语的含义。例如：

crop traits = traits exhibited by crops

注意等号两边名词的单复数形式。建议你总结读过论文中常见的“名词修饰名词”的情况，在学习它们时，尤其注意细节（单复数），这将大幅提高你论文写作的准确性。

如何阅读陌生科学领域的文章

综合 17.3 节和 17.4 节可知，科技文体包含大量句子模板和名词短语，句子模板通常可以再利用，名词短语则为具体的研究领域所特有。一旦理解了这一点，你或许会发现阅读不熟悉的科学领域的论文会变得更容易一些。初读这样的论文时，可以跳过陌生的名词短语，只抓住句子的框架，理解主要观点。然后判断哪些名词短语反复出现，并且有必要了解具体含义，利用词典或网络进行查询。需要查询多少名词短语跟你的阅读目的也有关。如果你作为初学者希望深入了解当前研究领域，那么你或许需要查询大量的名词短语；如果你只是希望通过阅读文章寻找某个特定信息（例如该研究使用的方法），那么查询的工作量就相应减少。名词短语的中心词与句意也有紧密关联，因此在确定是否需要查词典之前，可以先对中心名词进行分析。

掌握上述技能也可以帮助你更好地完成本书有关 PEA 的各类任务和习题，如果你不了解分子生物学和植物生理学，在阅读 PEA1 时，即可跳过复杂的名词短语，把注意力集中在句子结构上，这样的阅读方式或许能让你更顺利地完成相关习题任务，甚至还能让你更有效领会我们设置这些任务的初衷。同理，如果你选择了 PEA2，但是不了解海洋生物学，或者选择了 PEA3，但是并不了解火山或计算机建模，需要完成习题时，你都可以采用这种阅读方式。

17.5 专业英语的学习工具：语料库检索软件

每种语言都存在大量的词组和搭配，一些词通常就是需要跟另外一些词同时使用，例如 theory and practice，genetically modified organisms 以及 the effect of … on …等。这些搭配值得注意和学习。如果需要掌握你自己研究领域中的常用搭配，就得阅读这个领域内的论文，学习它们的语言。接下来，我们为你介绍一种软件，有了它的帮助，你可以更系统地学习和检索词语的用法。

语料库检索软件能做什么?

语料库检索软件可以从许多文本（语料库）中查询某个具体条目的所有用法，查询结果会列出包含这个条目的全部区段，每个区段占一行，该条目会出现在每行的中间并突出显示。查询结果有向左、向右两种分类（同时可以设定范围，例如距离该条目向左 3 个词），查询结果就可以作为语言学习的素材。如果语料库中的文本都与你的研究领域相关，那这个语料库就可以用来帮助你提升专业英语。

通过下面的任务 17.4，我们先来向你展示语料库检索结果可以揭示哪些写作知识。随后为你介绍如何从网络上下载免费的语料库检索软件 AdTAT，以及构建自己研究领域语料库的方法。

任务 17.4 如何观察检索结果

我们在土壤学的语料库中利用 AdTAT 检索了 soil 一词的用法，请查看检索结果，并阅读紧随其后的问题和回答。

to utilise existing available ***soil*** water, unlike the perennial gr
es (4 g oven dry wt basis) of ***soil*** were weighed into 40 ml polypr
required 9 kg P/ha, whereas a ***soil*** with a high P sorption capacit
concentration by 1 mg/kg on a ***soil*** with a low P sorption capacity
00, it was expected that this ***soil*** would have consistently been t
capacity (PBC), which is the ***soil***´s capacity to moderate changes
and buffering capacity of the ***soil***-an attempt to test Schofield´s
nisms that are present in the ***soil*** - plant microcosm environment. T
etermined in a growth-chamber ***soil*** - plant microcosm study. Nodding
84) Lime and phosphate in the ***soil*** - plant system. Advances in Agro
a where crops rely heavily on ***soil*** - stored water accrued in summer
fertility on these particular ***soil***s. Although this aberration has

over in a range of allophanic ***soil*** s amended with ^{14}C-labelled gluc

alues for 9 different pasture ***soil*** s, 6 and 12 months after P fert

问题 1：根据检索结果判断，soil 是可数名词还是不可数名词？

答：soil 既可用作可数名词，也可用作不可数名词，其中 a soil with a high P sorption capacity 和 9 different pasture soils 是可数的情况，samples of soil were weighed 是不可数的情况。

在日常写作中，soil 通常都不可数，可见科技文体中存在一些特殊用法。你也可以建立自己研究领域的语料库，利用 AdTAT 搜寻词语在专业英语中的特定用法。

问题 2：结合检索结果分析，soil 一词可以有几种用法？

答：能观察到许多用法。除了用作可数或不可数名词，soil 还出现在名词短语中，用作中心词（pasture soils），也可以修饰其他名词（soil water，soil－stored water，soil－plant microcosm）。

获取并使用 AdTAT 软件

AdTAT 是阿德莱德大学开发的基于 Java 的语料库检索工具，软件十分小巧，且操作简便，请前往 https：//www.adelaide.edu.au/carst/resources-tools/adtat/免费下载。借助 AdTAT，你可以迅速回答有关词语用法的疑难问题。

我们建议你利用自己学科的英文论文，建立一个语料库。这样可以满足你的个性化需求。把软件和语料库放在电脑桌面上，你就可以一边撰写论文，一边随时查证某些词组和表达在你的专业领域中具体的用法。

语料库的构建

由自己学科论文构成的语料库才最实用。如果你要发表综述类的文章，语料库也应该由综述构成。如果你要发表论文，那么可以精选 20～30 篇发表在某个研究领域内的论文，你还可以听取导师或研究团队负责人的建议，以确保：

- 论文来自领域内享有声望的期刊；
- 论文的语言质量很高，作者的母语是英语或具有同等水平；
- 论文覆盖了领域内不同的主题，可以提供丰富的语言素材；
- 文章类型与你的写作目标一致（例如语料库是否需要包含综述类文章）。

语料库软件可识别的文件类型

选入语料库的文章都应转化成 txt 格式，这样才能确保检索功能的正常使用。如果下载论文时，可以使用 html 格式预览全文，那么保存 txt 格式的文档就十分轻松了。如果论文作者愿意提供 word 版本的文件，转存成

txt也不存在困难。但是注意一点，生成的txt文件中需要删除所有图表、作者简介和参考文献列表。另一种情况是pdf格式的文档，转化成txt时需要多次重复同样的操作，详细步骤将在下文进行介绍。所有的txt文件都应存放在同一个文件夹内，以便于软件加载和检索。

版权问题

将某份电子版文档（txt）用于语料库检索软件就类似于在科研中使用来自数据库的论文（但是把内部构建的语料库文本对研究机构以外的人员开放可能会涉及侵权）。

软件操作说明

本书的网站上详细介绍了AdTAT软件的使用方法，你还可以参考软件内包含的帮助（Help）文件。

将pdf格式转换为txt文件

要多次进行复制、粘贴的操作，但只需复制论文的正文。所有的图表、作者简介、参考文献列表、致谢以及页眉页脚都不要复制。下面是详细步骤，可能需要多尝试几次，才能很熟练地完成转换。

- 下载论文（网络上的pdf文件）；
- 使用Adobe Acrobat Reader打开文件；
- 在不选中页眉、页脚、页码、图表和参考文献的前提下，尽量多地选取文件的正文；
- 将文本复制（Ctrl+C）；
- 打开word等文字处理软件；
- 将文本粘贴到新建的文件中（Ctrl+V）；
- 反复进行选择、复制、粘贴的步骤，直至把所有必要的内容都保存下来；
- 在word软件的功能区选择“文件”菜单内的“另存为”；
- 在弹窗中选择文件类型为txt，并在保存前修改文件名；
- 可以视实际需要对txt文件进一步编辑（删掉多余内容）。

一些pdf文件进行了加密处理，无法直接复制。此时，复制粘贴命令是失效的，那我们也无计可施。把复制好的内容存入新文件时，一定注意不要包含页眉、页脚和页码。根据我们的经验，最高效的做法是先复制整页或整个一栏文字，粘贴后再对文本进行编辑（删去多余的空格、回车等），将断断续续的内容复原。这样操作可以从一开始就避免选中不必要的内容。整个转换和编辑过程比较枯燥，但熟练后就变成了机械化的体力劳动。

请借助AdTAT完成任务17.5。

任务 17.5 使用 AdTAT 进行检索

利用 AdTAT 软件（或仔细阅读语料库中的论文），寻找下列问题的答案：

1. 论文作者是否在句子的开头使用 Also?
2. 论文作者是否在句子的开头使用 In addition?
3. addition 还有什么样的用法?
4. 论文作者是否使用了 I 或 we?
5. 动词 affect 常用于什么样的结构?
6. 哪些动词可以与名词 role 搭配? 在 role 后面经常使用哪些介词?

你还可以进行哪方面的检索? 我们在网站上（http://www.writeresearch.com.au）提供了更多的检索思路。

17.6 恰当使用冠词(a/an, the)

对于母语非英语的作者而言，由冠词引发的困惑从初学阶段就一直存在。你或许也看过许多不同的说法，并且反复学习过它们。我们在这里再次讨论冠词，主要原因还是由于冠词难以掌握，尤其是针对那些在母语中不使用冠词的作者。同时，编辑和审稿人也十分看重英文论文中冠词的使用。

事实上，如果编辑和审稿人的母语是英语，掌握冠词主要是通过沉浸在母语环境中，通过聆听家人的一言一语养成正确的使用习惯，因此这些人反而难以理解冠词的复杂性。据我们所知，市面上没有任何一款软件能够有效校对冠词的误用，这既说明冠词确实复杂，也体现出冠词的选用是与语境高度相关的，尤其是在判断特指还是泛指的时候。在写作中，选错冠词也会对语义产生影响，下面我们就先来分析特指与泛指。

表示泛指的名词短语

所谓泛指（generic noun phrase），代表的就是某类物体、概念或生物体中的任一成员或所有成员。用英语表示泛指有如下四种方式。

1. 如果名词可数，那么就用其复数形式表示泛指，名词前不需要加冠词。

2. 如果名词可数，还可以用 a/an 加上其单数形式表示泛指。例如：

Healthy crops can contribute substantial cadmium to human diets.

A healthy crop can contribute substantial cadmium to human diets.

3. 如果名词不可数，直接使用该名词（注意：用作不可数名词时不存在复数形式），前面不需要加冠词就表示泛指。例如：

Cadmium exists in soils in many forms.

Manipulation of soil pH can be effective in managing Cd contamination.

4. 还有一种情况值得注意：英语中也可以使用定冠词 the 加上可数名

词单数表示泛指的概念。此类名词通常是有生命的或是某种常用机械或设备（也允许替换成上面的第一种情况，使用名词的复数，但记得调整动词形式，做到主谓一致）。例如：

The earthworm can be found in many types of soil.（或 Earthworms can...）

The computer has become an important tool for researchers.（或 Computers have...）

注意：在科技文体中，即使你再次使用某个名词，只要你讨论的是泛指的概念或一般的类别，这个名词就不应该理解成特指（也就是需要你忘记如下规则：如果某个名词在上文中出现过一次，下次出现就代表特指）。请完成任务 17.6。

任务 17.6 表示泛指的名词短语

下面的段落出自 PEA 1 的引言部分，请用下划线标出表示泛指的名词短语。

Legumes form symbiotic associations with N_2-fixing soil-borne bacteria of the *Rhizobium* family. The symbiosis begins when compatible bacteria invade legume root hairs, signalling the division of inner cortical root cells and the formation of a nodule. Invading bacteria migrate to the developing nodule by way of an 'infection thread', comprised of an invaginated cell wall. In the inner cortex, bacteria are released into the cell cytosol, enveloped in a modified plasma membrane (the peribacteroid membrane (PBM)), to form an organelle-like structure called the symbiosome, which consists of bacteroid, PBM and the intervening peribacteroid space (PBS; Whitehead and Day, 1997). The bacteria, subsequently, differentiate into the N_2-fixing bacteroid form. The symbiosis allows the access of legumes to atmospheric N_2, which is reduced to NH_4^+ by the bacteroid enzyme nitrogenase. In exchange for reduced N, the plant provides carbon to the nodules to support bacterial respiration, a low-oxygen environment in the nodule suitable for bacteroid nitrogenase activity, and all the essential nutritional elements necessary for bacteroid activity. Consequently, nutrient transport across the PBM is an important control mechanism in the promotion and regulation of the symbiosis.

请核对书后的参考答案。

表示特指的名词短语

所谓特指（specific noun phrase），代表的就是某类事物中的具体、特定个体，而不是表示该类别的整体。读者和论文作者都知道名词指代的是哪些或是哪个个体，此时就需要定冠词 the 表示特指。请体会以下三种情况。

1. 如果名词词组的指代对象（referent）是读者和论文作者共知的信息，用特指。例如：

In recent years the growth of desert areas has been accelerating in the world.

2. 如果名词词组指代了向读者介绍过的旧信息，用特指。例如：

A pot experiment was conducted in an acid soil. The experiment showed...

3. 如果名词词组后面紧跟了修饰成分说明具体说明它（它们）具体指的是哪一个或哪一些，用特指。例如：

The aim of this study was to investigate the effect of liming on Cd uptake.

注意：如果使用 NP1+of+NP2 的结构，位于前面的名词短语是特指（加 the）的可能性高达 85%，因此在这种情况下建议你使用定冠词 the，除非你十分确信 NP1+of+NP2 这个整体代表的概念是一般性、非特指的。请完成任务 17.7。

任务 17.7 表示特指的名词短语

再次阅读任务 17.6 中出自 PEA1 引言部分的段落，请用方框圈出表示特指的名词短语。如果可能的话，与同事讨论它们为何表示特指概念。请核对书后的参考答案。

选定冠词的流程

图 17.1 是选用冠词时可以使用的流程图，它将帮助母语非英语的作者判断 a，an 或 the 到底该如何用在具体的语句中。请完成任务 17.8。

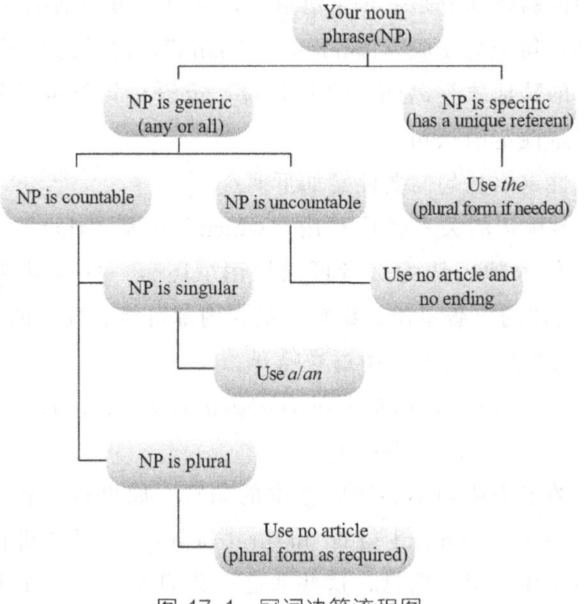

图 17.1 冠词决策流程图

任务 17.8 科技文体中的冠词和名词复数

阅读下面的段落，借助图 17.1，在必要处填入复数 - s 或冠词 a，an，the。（有些位置什么都不需要填）

Propagule pressure

____propagule pressure is widely recognized as____important factor that influences____invasion success. ____previous studies suggest that____probability of____successful invasion increases with____number of propagules released, with____number of introduction attempts, with ____ introduction rate, and with ____ proximity to ____ existing populations of invaders. Moreover, ____propagule pressure may influence____invasion dynamics after____establishment by affecting____capacity of____non-native species to adapt to their new environment. Despite its acknowledged importance, ____propagule pressure has rarely been manipulated experimentally and____interaction of____propagule pressure with____other processes that regulate____invasion success is not well understood.

请核对书后的参考答案。

17.7 正确使用“which”和“that”

有关“which”和“that”的疑问是二者的区别，以及到底是否要使用逗号。我们在编辑科技论文时，经常会看到相关的错误。希望通过下面的说明，你能理解“which”和“that”的使用规则。

例句 1：Land *which is surrounded by water* is an island.

例句 1 中的斜体部分是一个定语从句，它对于整句话的含义至关重要。如果将其省略，句子会变成“Land is an island”，这样的句子是说不通的，因为四面环水的陆地才是岛屿。也就是说，定语从句限定了陆地的具体所指，是一个限定性定语从句。

使用限定性定语从句需要注意如下两点。

- 引导定语从句的关系代词是用“which”还是“that”没有定论。一些作者（和写作教师）认为二者可以换用。还有一些人认为应统一使用“that”。对于具体的出版单位，编辑人员在内部可能有统一的规则。
- 定语从句和主句之间不用逗号隔开。

例句 2：Tasmania，*which is surrounded by the waters of Bass Strait*，is an island of great natural beauty.

例句 2 中的定语从句不会影响基本的句意。如果将其省略，句子会变成“Tasmania is an island of great natural beauty”，仍然讲得通。也就是说，定语从句在补充额外信息，这些非必需信息构成了一个非限定性定语从句。鉴别非限定性定语从句时，可以在“which”后面插入“by the way”，看看句意是否说得通。

使用非限定性定语从句需要注意如下两点。

- 定语从句和主句之间用逗号隔开。如果定语从句出现在句子中间，那么前后都要使用逗号（见例句 2）；如果出现在句尾，只需在定语从句之前加一个逗号。
- 引导定语从句的关系代词只能是 which。

注意：如果从句使用了缩略式，去掉主谓语（which/that＋谓语动词），保留分词短语，那么是否使用逗号依然可以参考上述规则。例如：

Tasmania，surrounded by the waters of Bass Strait，is an island of great natural beauty.

Land surrounded by water is an island.

请完成任务 17.9。

任务 17.9　添加标点符号

请为下面的句子加上标点符号。

1. Lime which raises the pH of the soil to a level more suitable for crops is injected into the soil using a pneumatic injector.
2. Manipulation which involves adding or deleting genetic information is referred to as genetic engineering.
3. Non-cereal phases which are essential for the improvement of soil fertility break disease cycles and replace important soil nutrients.
4. Senescence which is the aging of plant parts is caused by ethylene that the plant produces.
5. Opportunities that arise from the economically buoyant nature of domestic wine production must be identified and carefully assessed.
6. Seasonal cracking which is a notable feature of this soil type provides pathways at least 6 mm wide and 30 cm deep that assist in water movement into the subsoil.
7. Plants which experience waterlogging early in their development would be expected to have a much shallower root system than non-waterlogged plants.
8. Yellow lupin which may tolerate waterlogging better than the narrow-leafed variety has the potential to improve yields in this area.
9. Lucerne is a drought-hardy perennial legume which produces high-quality forage.

请核对书后的参考答案。

18. 科研项目申请书

18.1　准备工作

在科研生涯中，科研经费扮演着重要角色。在大多数情况下，申请人获取科研经费需要结合课题申报范围和要求提交申请书，论证课题设计，并通过资助单位的审批。整个过程竞争异常激烈，大多数申请书都经过精心准备，但最终成功立项的数量却十分有限。这与论文作者向高水平期刊投稿的情况极为类似。借助前面章节介绍过的写作原则和策略，本章将聚焦科研项目申请书，但是我们不会讨论每一项评价指标，尤其不会涉及课题内容的科学性。在前面的章节中，本书汇集了众多研究者在针对科技论文进行语类分析（也称体裁分析）时得出的结论，但是还没有研究者针对科研项目申请书这一语类进行分析，主要的原因可能是收集成功的案例素材比较困难，另一方面，科研项目申请书需要满足的要求也更加五花八门。我们在撰写这一章时，借鉴了论文写作过程中值得吸取的经验教训，还邀请其他作者和具有不同专业背景的科研人员，也就是我们眼中有鉴赏力的读者试读了本章的草稿，并参考了相关反馈意见。因此我们相信，即将提供给你的建议能帮助你精简写作流程，写出令人信服的科研项目申请书。

撰写科研项目申请书同样极度耗时耗力，掌握好时机至关重要。首先就需要问自己：现在是申请课题的适宜时机吗？到底是应该先完成写到一半的论文，还是有信心利用额外的时间完成一份高质量的申请？另一项值得考虑的因素是自己开展过的研究工作，已有的研究结果能带来哪些灵感，能够为此次的选题提供支撑吗？为了更好地应对时间压力，平时就要注意对在研项目进行归档，随时充实相关细节，一旦项目申报开启，确保能迅速组织材料。平时需要记录和归档的内容因人而异，你可以继续阅读下面的写作步骤，标记出平时要积累的信息类型。

不同的资助单位对申请人提出的要求各不相同，申请流程和审批标准也并不统一，因此提醒你千万不要盲目遵循下面的建议，要结合自己的情况判断是否适用。申请书的类型、格式，申请流程和审批标准也受到项目所属的学科分支，以及国家和文化背景等因素的影响。但是对于任何一份科研项目申请书而言，必须满足以下两个条件：一是申请的课题具有科学性、可行性，适合在当前时期开展；二是申请人遵守课题指南的要求，提供了所有重要信息（剔除了任何多余的信息）。拿出资助单位提供的资助方案，找到和自己研究领域相关的具体要求，这值得你花时间仔细阅读。

18.2 写作策略与步骤

科研项目申请书起草过程中值得关注的五个问题

在本书的第 4 章，我们提出了四个论文起草过程中值得关注的问题。针对申请书的写作准备阶段，请思考下列五个问题。你的答案将帮助你规划申请流程，选定合适的资助基金。

- 该科研项目要达到的目标是什么（具体、可量化的结果）？
- 在现阶段，这些研究结果的重要性是什么？
- 项目团队需要哪些成员加入，以确保项目如期高质量完成？
- 哪些资助团体或机构可能对该科研项目感兴趣？
- 为何要选择资助你的项目而不是其他申请人的项目？

将科研项目与资助单位进行匹配

此时需要考虑一系列问题，诸如：

- 项目是何种类型（例如：理论研究还是应用研究）？
- 科研项目要解决的问题是什么？
- 项目有多大的价值和影响力，能否实现双赢（例如：即使研究结果不显著，结论依然有价值）？
- 项目的规模与资助的时限是否匹配？
- 资助单位如何看待团队合作项目？对参与团队的属性（国内、国际、跨学科）有何偏好？在论证当前项目时，团队以往业绩所占的权重是多少？

在决策时，要仔细分析资助单位提供的所有文件；如果可能，还应分析获批的项目课题，以及这些项目的目标和结题情况。查看基金的宣传材料或资助单位的年报，找出它们关注的其他因素，例如研究结果发表在哪些期刊上，在何处宣传推广，以及年度报告书中侧重强调了资助项目的哪些方面。在申请书中从类似的角度描述自己的研究课题，有助于获得资助单位的青睐。

一旦选定了资助单位，为了更高效地争取到本次以及未来的资助机会，接下来就要梳理申请流程，记录重要的时间节点。主要的流程有：审阅资助标准和要求；了解资助单位对你所在科研机构的要求和许可情况；研究团队沟通和组建；课题设计前期准备；按提纲分段进行写作，包括（负责人）明确各项分工、汇总和校对；同事交流、互审；最终提交。

确定申请书要表达的关键信息

准备这一部分时，只需要完成课题名称和一段项目概述。我们建议还是先撰写项目的目标，这是整段概述在结尾处需要强调的内容。然后补充关键背景信息，也就是相关的介绍性话语，在这个过程中，把资助单位的关注点与项目要填补的研究空白连接起来，论证该课题的重要性。整段概述可以借用论文引言部分的论述模式（见第 8 章），突出层级 4（研究目标，写作的第一步）、层级 3（指出存在的问题、研究空白或进一步研究的必要性）和层级 1（根据资助单位的偏好阐释课题的价值）的相关内容。等到申请书全部完成后，再次审阅选题依据、思路方法和预期成果部分的内容，看是否与概述中的研究目标完全一致。课题名称可以随后敲定，我们在第 10 章中提供的建议同样适用。

接下来，根据资助单位提供的申请书格式，将预备写入各个主要部分的关键信息逐条列出。在列举这些信息时，确保能突出那些高度符合资助标准和要求的内容。做好这些后，你就可以准备邀请同事、同行对你的申请书提供反馈意见了。

第一轮论证: 征集导师和同事的意见

准备好你在上一步完成的申请书大纲，邀请当前研究领域资深的研究者，尤其是接受过该基金资助的研究者提出反馈意见。注意向前辈询问下列问题：申请书是否符合资助单位的偏好，以及在下一阶段撰写整份申请书的注意事项。

收到反馈意见后，先分析能否在合理的时间内得到满意的解决。如果时间和能力都允许，就可以制定改进方案，听取团队成员的建议，设定关键时间节点，要在最终提交申请书之前留出正式内部评审的时间。在调整时，可以参考我们在第二步“将科研项目与资助单位进行匹配”列出的主要流程，根据实际情况进行优化。

分析审批标准

与团队成员一起讨论资助单位的审批标准。如果能找到相关数据，还应考虑不同标准所占的权重。分析申请书将如何满足资助单位规定的标准，需要提供哪些证据，分别写入申请书的哪个部分。记录暂时无法提供的信息，确保在后期能够把它们补充到相应的部分中。

汇总并修改每个部分, 完成申请书初稿

将团队成员各自负责的部分汇总成完整的申请书，检查是否符合课题指南的要求，以及是否包含了针对评估标准所搜集的全部论据。

客观分析自己的课题，列出潜在的困难和缺陷。如果你能预想到相关问题，那么资助单位的评审专家也很可能想得到。确保在科研项目申请书中阐明自己将如何克服或应对这些困难。

在团队内部修改申请书时，需要特别注意科技文体的逻辑性（见第 8 章），理顺各个部分之间的论证逻辑。这是申请书的目标读者，也就是评审专家尤其看重的。评审专家未必精通你的研究主题，一些国家级科学基金项目的首轮评审尤其如此，因而这些读者很可能无法在申请书的各部分之间建立联系，除非你借助写作策略，在文字中清晰明确地把这些联系呈现

出来。资助单位通常对申请书的字数或页数有明确要求，切记不要超出限制。

第二轮论证：正式内部评审

及时开展第二轮反馈，最好能邀请到接受过该基金资助的研究者或为该资助机构担任过评审专家的人员为你审稿。正式内部评审可以模拟申请书提交后的筛选流程，审稿目的是找出项目的错漏之处和薄弱环节，以便在最终提交前再次修改。你所在的科研机构不一定会组织这一轮评审，但你可以邀请同事站在资助单位的角度对申请书做出评价。如果有明确的评审标准，记得提供给你的同事；如果没有具体的标准，可以请同事回答下面两个问题。

- 你是否愿意出资支持这个课题？
- 如果不愿意，申请人需要如何修改申请书，增加胜算？

另外，还可以专门寻求语言方面的反馈。请为你审稿的人评价申请书是否易于理解；为了弄清某些信息，是否需要反复阅读某段内容？收到反馈意见后，需要考虑怎样进一步修改申请书。但是注意，提供反馈意见的人可能会误解申请书的内容，提出不恰当的建议，毕竟这些人不是你研究领域内的专家。你需要有选择性地听取相关意见。

为下一次申请做准备

无论成功与否，你都需要记下资助单位给出的反馈意见，并把这些意见融入你未来的写作策略中。结合具体的反馈，分析可以在准备阶段的哪个时期做出调整，避免下一次申请时再次犯错。为确保下一次项目申报开启时，依然能迅速组织材料，应该保持更新以下内容：

- 在研项目的日常归档，随时充实相关细节（见 18.1 节）；
- 不断优化项目申请流程和时间节点（第二步“将科研项目与资助单位进行匹配”）。

时刻关注所在科研机构的相关信息，主动获取指导和帮助，学习写出更有说服力的申请书，并及时了解最新的资助机会。Johnson（2011）发表在“Elsevier”上的白皮书中也提供了申报科研项目的建议，可以到网络上免费下载、阅读（第十章：Preparing a Grant Funding Application）。

18.3 容易出现的错误

申请失败的主要原因就是准备仓促或不充分，下面列出一些常见错误，虽不全面，但足以说明问题。

- 忽略课题指南的要求，未按照标准提供相应内容或内容的数量不足。资助单位要面对大量的申请人，自然希望收到高质量、符合要求的申请书，这体现了申请人的专业素养。

• 未充分理解资助者的偏好、目标和资助范围。申请人的成败由资助单位决定。

• 目标读者（评审专家）在理解申请书不同部分的语言时存在困难。例如，选题依据（引言）部分的读者可能不是领域内的专家，而方法部分的读者可能是某学科或分支学科的专家。

• 项目列出的时间节点、研究范围或经费概算不切实际。资助者往往有能力判断申请人能否在既定时间内兑现相应的承诺。

• 花太多笔墨介绍研究活动的细节，但弱化了其意义和价值。资助单位希望投资成功，看重项目的价值。

• 在描述项目课题时，未明确呼应资助者的标准和关心的问题。评审专家不是在鉴赏散文，没时间深挖你隐藏在字里行间的关键信息。

• 申请书中包含了无用的行话、术语、套话、缩略词，或行文冗长、不够简洁。这些内容极度影响了沟通的清晰性、准确性。

• 申请书用词不精确，存在拼写或语法错误，或使用了错误的符号、标记。务必认真校对。草率的研究者才会写出草率的申请书。

在准备和修改申请书时，建议你反复核查上述错误。这样做会使你站在目标读者，也就是资助者的角度思考，从最大程度上满足资助单位的要求。

05 第五部分

论文示例

CHAPTER 19

PEA1：Kaiser et al. （2003）

The Plant Journal (2003) **35**, 295–304 doi: 10.1046/j.1365-313X.2003.01802.x

The soybean NRAMP homologue, GmDMT1, is a symbiotic divalent metal transporter capable of ferrous iron transport

Brent N. Kaiser[1], Sophie Moreau[2], Joanne Castelli[3], Rowena Thomson[3], Annie Lambert[2], Stéphanie Bogliolo[4], Alain Puppo[2] and David A. Day[3,*]

[1]*School of Agricultural Sciences, Discipline of Wine & Horticulture, The University of Adelaide, Urrbrae, South Australia, Australia,*

[2]*Laboratoire de Biologie Végétale et Microbiologie, CNRS FRE 2294, Université de Nice-Sophia Antipolis, Parc Valrose, 06108 Nice cédex 2, France,*

[3]*Biochemistry & Molecular Biology, School of Biomedical & Chemical Sciences, University of Western Australia, Crawley, WA 6009, Australia, and*

[4]*Laboratoire de Physiologie des Membranes Cellulaires, UMR 6078 CNRS-Université de Nice-Sophia Antipolis, 284 chemin du Lazaret, 06230 Villefranche sur Mer, France*

Received 9 December 2002; revised 24 April 2003; accepted 7 May 2003.
*For correspondence (fax +61 08 9380 1148; e-mail dday@cyllene.uwa.edu.au).

Summary

Iron is an important nutrient in N_2-fixing legume root nodules. Iron supplied to the nodule is used by the plant for the synthesis of leghemoglobin, while in the bacteroid fraction, it is used as an essential cofactor for the bacterial N_2-fixing enzyme, nitrogenase, and iron-containing proteins of the electron transport chain. The supply of iron to the bacteroids requires initial transport across the plant-derived peribacteroid membrane, which physically separates bacteroids from the infected plant cell cytosol. In this study, we have identified *Glycine max divalent metal transporter 1* (*GmDmt1*), a soybean homologue of the NRAMP/Dmt1 family of divalent metal ion transporters. *GmDmt1* shows enhanced expression in soybean root nodules and is most highly expressed at the onset of nitrogen fixation in developing nodules. Antibodies raised against a partial fragment of GmDmt1 confirmed its presence on the peribacteroid membrane (PBM) of soybean root nodules. GmDmt1 was able to both rescue growth and enhance ^{55}Fe(II) uptake in the ferrous iron transport deficient yeast strain (*fet3fet4*). The results indicate that GmDmt1 is a nodule-enhanced transporter capable of ferrous iron transport across the PBM of soybean root nodules. Its role in nodule iron homeostasis to support bacterial nitrogen fixation is discussed.

Keywords: iron, NRAMP, nitrogen fixation, soybean, symbiosome.

Introduction

Legumes form symbiotic associations with N_2-fixing soil-borne bacteria of the *Rhizobium* family. The symbiosis begins when compatible bacteria invade legume root hairs, signalling the division of inner cortical root cells and the formation of a nodule. Invading bacteria migrate to the developing nodule by way of an 'infection thread', comprised of an invaginated cell wall. In the inner cortex, bacteria are released into the cell cytosol, enveloped in a modified plasma membrane (the peribacteroid membrane (PBM)), to form an organelle-like structure called the symbiosome, which consists of bacteroid, PBM and the intervening peribacteroid space (PBS; Whitehead and Day, 1997). The bacteria, subsequently, differentiate into the N_2-fixing bacteroid form. The symbiosis allows the access of legumes to atmospheric N_2, which is reduced to NH_4^+ by the bacteroid enzyme nitrogenase. In exchange for reduced N, the plant provides carbon to the nodules to support bacterial respiration, a low-oxygen environment in the nodule suitable for bacteroid nitrogenase activity, and all the essential nutritional elements necessary for bacteroid activity. Consequently, nutrient transport across the PBM is an important control mechanism in the promotion and regulation of the symbiosis.

Micronutrients such as iron are essential for bacteroid activity and nodule development. The demand for iron increases during symbiosis (Tang *et al.*, 1990), where the

(a)

```
                                     taataataaagctaaatca  -77
tagtagtgaggagtgactagtacaaacagaatccaaagcttttttttttc  -27
ttcttcttcttcttctttctaacgccATGTCTGGGAGCCACCAAGAGCAG   24
                             M  S  G  S  H  Q  E  Q    8
CCACTGTTAGAGAACTCGTTCATAGAAGAAGACGAGCCGCAAGAAACAGC   74
 P  L  L  E  N  S  F  I  E  E  D  E  P  Q  E  T  A   25
TTATGATTCGTCGGAGAAGATAGTGGTGGTCGGAGTCGACGAGTTCGATG  124
  Y  D  S  S  E  K  I  V  V  V  G  V  D  E  F  D     41
ACGAGGAGAATTGGGGGAGAGTGCCCCGATTCTCGTGGAAGAAGCTATGG  174
D  E  E  N  W  G  R  V  P  R  F  S  W  K  K  L  W   58
CTGTTCACCGGGCCGGGCTTTCTGATGAGCATAGCGTTTCTGGACCCTGG  224
 L  F  T  G  P  G  F  L  M  S  I  A  F  L  D  P  G   75
AAACTTAGAGGGGGACCTTCAGGCGGGTGCCATTGCAGGGTACTCATTGT  274
  N  L  E  G  D  L  Q  A  G  A  I  A  G  Y  S  L     91
TGTGGCTTCTGATGTGGGCCACAGCAATGGGCCTCCTGATCCAGCTCCTC  324
L  W  L  L  M  W  A  T  A  M  G  L  L  I  Q  L  L   108
TCGGCACGGCTCGGCGTGGCCACAGGGAAGCACCTCGCCGAGCTCTGCCG  374
 S  A  R  L  G  V  A  T  G  K  H  L  A  E  L  C  R  125
AGAGGAGTATCCTCCGTGGGCCCGGATAGTGCTCTGGATCATGGCGGAAC  424
  E  E  Y  P  P  W  A  R  I  V  L  W  I  M  A  E    141
TCGCTCTCATTGGCTCCGATATTCAGGAGGTTATTGGGAGCGCTATTGCA  474
L  A  L  I  G  S  D  I  Q  E  V  I  G  S  A  I  A   158
ATCAGGATTCTTAGTCATGGGGTTGTGCCCCTTTGGGCTGGGGTTGTCAT  524
 I  R  I  L  S  H  G  V  V  P  L  W  A  G  V  V  I  175
TACTGCTCTTGATTGTTTTATTTTTCTCTTTCTTGAGAACTATGGTGTGA  574
  T  A  L  D  C  F  I  F  L  F  L  E  N  Y  G  V    191
GGACTTTGGAAGCTTTTTTTGCTATTCTCATTGGTGTGATGGCAATCTCG  624
R  T  L  E  A  F  F  A  I  L  I  G  V  M  A  I  S   208
TTCGCATGGATGTTTGGTGAAGCCAAGCCCAGTGGCAAGGAACTTCTTCT  674
 F  A  W  M  F  G  E  A  K  P  S  G  K  E  L  L  L  225
TGGAGTTTTGATTCCAAAACTCAGCTCCAAAACTATACAGCAGGCTGTTG  724
  G  V  L  I  P  K  L  S  S  K  T  I  Q  Q  A  V    241
GAGTTGTGGGGTGCCTTATTATGCCTCACAATGTGTTCTTGCACTCTGCT  774
G  V  V  G  C  L  I  M  P  H  N  V  F  L  H  S  A   258
CTTGTTCAGTCAAGGCAGGTTGACCGCAGCAAGAAAGGCCGAGTTCAAGA  824
 L  V  Q  S  R  Q  V  D  R  S  K  K  G  R  V  Q  E  275
AGCTCTTAATTATTACTCGATAGAGTCCACCCTTGCCCTTGTAGTTTCCT  874
  A  L  N  Y  Y  S  I  E  S  T  L  A  L  V  V  S    291
TTATTATAAATATTTTTGTAACAACAGTGTTTGCTAAGGGATTTTATGGC  924
F  I  I  N  I  F  V  T  T  V  F  A  K  G  F  Y  G   308
TCTGAACTTGCAAACAGCATAGGTCTTGTAAATGCAGGACAGTATCTAGA  974
 S  E  L  A  N  S  I  G  L  V  N  A  G  Q  Y  L  E  325
GGAGACATATGGGGGTGGACTATTTCCAATTTTATACATATGGGGTATTG 1024
  E  T  Y  G  G  G  L  F  P  I  L  Y  I  W  G  I    341
GATTATTAGCAGCAGGCCAAAGTAGCACTATTACTGGGACTTATGCAGGA 1074
G  L  L  A  A  G  Q  S  S  T  I  T  G  T  Y  A  G   358
CAATTCATCATGGGAGGTTTTCTAAATTTAAGGTTAAAGAAGTGGATGAG 1124
 Q  F  I  M  G  G  F  L  N  L  R  L  K  K  W  M  R  375
GGCGTTGATTACCCGAAGTTGTGCAATAATTCCAACTATGATAGTTGCTC 1174
  A  L  I  T  R  S  C  A  I  I  P  T  M  I  V  A    391
TTTTATTCGATACCTCGGAGGAATCGTTAGATGTTTTGAATGAGTGGCTT 1224
L  L  F  D  T  S  E  E  S  L  D  V  L  N  E  W  L   408
AATGTTCTTCAGTCAGTCCAGATCCCCTTTGCACTTATTCCCTTGCTTTG 1274
 N  V  L  Q  S  V  Q  I  P  F  A  L  I  P  L  L  C  425
TCTGGTGTCAAAGGAGCAGATAATGGGCACTTTCAGAATTGGTGCTGTCC 1324
  L  V  S  K  E  Q  I  M  G  T  F  R  I  G  A  V    441
TCAAGACTACTTCATGGCTCGTGGCTGCTCTGGTGATAGTGATTAATGGC 1374
L  K  T  T  S  W  L  V  A  A  L  V  I  V  I  N  G   458
TATCTTTTGACGGAATTCTTTTCCTCTGAAGTGAATGGACCAATGATTGG 1424
 Y  L  L  T  E  F  F  S  S  E  V  N  G  P  M  I  G  475
CACTGTAGTGGGTGTAATAACTGCTGCATATGTTGCCTTCGTAGTATACC 1474
  T  V  V  G  V  I  T  A  A  Y  V  A  F  V  V  Y    491
TTATTTGGCAAGCCATCACCTATTTACCTTGGCAAAGTGTAACACAACCA 1524
L  I  W  Q  A  I  T  Y  L  P  W  Q  S  V  T  Q  P   508
AAGACAATTGCTCATTCAGAGGGTTGAggttgagtgatcaatctttaaaa 1574
 K  T  I  A  H  S  E  G  *                           516
tcgcggaataggaagtgccatccatttttaagtatgctcatgcttgtttg 1624
ttactcgtgtggcaagttgatgcaataggtggtggcaccttattctttgc 1674
ctgtaattataaactatgtcagagtagattttagctctgtattagtactt 1724
tcaaattttgttgtcaaaaaaaaaaaaaaa                     1754
```

(b)

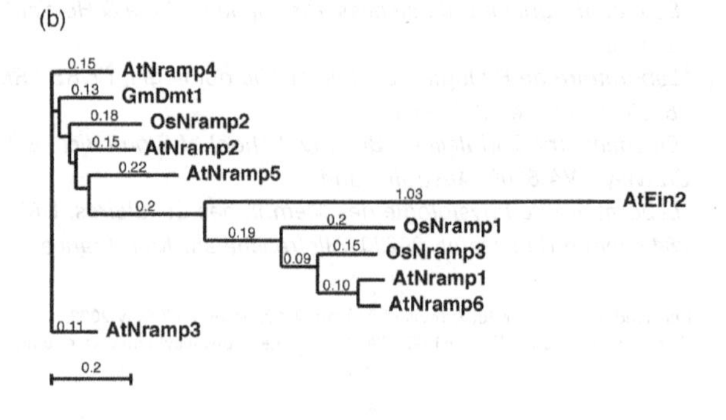

(c)

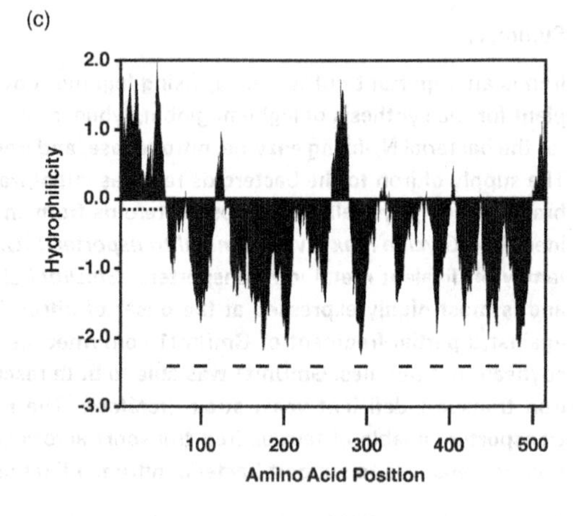

metal is utilised for the synthesis of various iron-containing proteins in both the plant and the bacteroids. In the plant fraction, iron is an important part of the heme moiety of leghemoglobin, which facilitates the diffusion of O_2 to the symbiosomes in the infected cell cytosol (Appleby, 1984). In bacteroids, there are many iron-containing proteins involved in N_2 fixation, including nitrogenase itself and cytochromes used in the bacteroid electron-transport chain. In the soil, iron is often poorly available to plants as it is usually in its oxidised form Fe(III), which is highly insoluble at neutral and basic pH. To compensate this, plants have developed two general strategies to gain access to iron from their localised environment. Strategy I involves secretion of phytosiderophores that aid in the solubilisation and uptake of Fe(III), while strategy II involves initial reduction of Fe(III) to Fe(II) by a plasma membrane Fe(III)-chelate reductase, followed by uptake of Fe(II) (Romheld, 1987). The mechanism(s) involved in bacteroid iron acquisition within the nodule have been investigated at the biochemical level, and three activities have been identified (Day *et al.*, 2001). Fe(III) is transported across the PBM complexed with organic acids such as citrate, and accumulates in the PBS (Levier *et al.*, 1996; Moreau *et al.*, 1995), where it becomes bound to siderophore-like compounds (Wittenberg *et al.*, 1996). Fe(III) chelate reductase activity has been measured on isolated PBM, and Fe(III) uptake into isolated symbiosomes is stimulated by Nicotinamide Adenine Dinucleotide (NADH), reduced form (Levier *et al.*, 1996). However, Fe(II) is also readily transported across the PBM and has been found to be the favoured form of iron taken up by bacteroids (Moreau *et al.*, 1998). The proteins involved in this transport have not yet been identified.

Two classes of putative Fe(II)-transport proteins (Irt/Zip and Dmt/Nramp) have been identified in plants (Belouchi *et al.*, 1997; Curie *et al.*, 2000; Eide *et al.*, 1996; Thomine *et al.*, 2000). The Irt/Zip family was first identified in *Arabidopsis* by functional complementation of the yeast Fe(II) transport mutant DEY1453 (*fet3fet4*; Eide *et al.*, 1996). *AtIrt1* expression is enhanced in roots when grown on low iron (Eide *et al.*, 1996), and appears to be the main avenue for iron acquisition in *Arabidopsis* (Vert et *al.*, 2002). Recently, a soybean Irt/Zip isologue, GmZip1, was identified and localised to the PBM in nodules (Moreau *et al.*, 2002). GmZip1 has been characterised as a symbiotic zinc transporter, which does not transport Fe(II). The second class of iron-transport proteins consists of the Dmt/Nramp family of membrane transporters, which were first identified in mammals as a putative defence mechanism utilised by macrophages against mycobacterium infection (Supek *et al.*, 1996; Vidal and Gros, 1994). Mutations in Nramp proteins in different organisms result in varied phenotypes including altered taste patterns in *Drosophila* (Rodrigues *et al.*, 1995), microcytic anaemia (mk) in mice and belgrade rats (Fleming *et al.*, 1997) and loss of ethylene sensitivity in plants (Alonso *et al.*, 1999). The rat and yeast NRAMP homologues (DCT1 and SMF1, respectively) have been expressed in *Xenopus* oocytes and shown to be broad-specificity metal ion transporters capable of Fe(II), amongst other divalent cations, transport (Chen *et al.*, 1999; Gunshin *et al.*, 1997). The plant homologue, AtNramp1, complements the growth defect of the yeast Fe(II) transport mutant DEY1453, while other *Arabidopsis* members do not (Curie *et al.*, 2000; Thomine *et al.*, 2000). Interestingly, AtNramp1 overexpression in *Arabidopsis* also confers tolerance to toxic concentrations of external Fe(II) (Curie *et al.*, 2000), suggesting, perhaps, that it is localised intracellularly.

In this study, we have identified a soybean homologue of the Nramp family of membrane proteins, GmDmt1;1. We show that GmDmt1;1 is a symbiotically enhanced plant protein, expressed in soybean nodules at the onset of nitrogen fixation, and is localised to the PBM. GmDmt1;1 is capable of Fe(II) transport when expressed in yeast. Together, the localisation and demonstrated activity of GmDmt1;1 in soybean nodules suggests that the protein is involved in Fe(II) transport and iron homeostasis in the nodule to support symbiotic N_2 fixation.

Results

Cloning of GmDmt1;1

A partial cDNA of GmDmt1;1 was identified from a 6-week-old soybean nodule cDNA library during a 5′-RACE PCR experiment designed to amplify the N-terminal sequence of a putative NH_4^+ transporter, GmAMT1. Subsequent PCR experiments identified a full-length 1849-bp cDNA, which was cloned and sequenced (Figure 1a) (accession no.

Figure 1. Sequence analysis.
(a) Nucleotide and the deduced amino acid sequence of *GmDmt1;1*. Amino acids italicised and in bold represent the N-terminal region of GmDmt1;1 used for the generation of the anti-GmDmt1;1 antisera. Consensus Dmt transport motif (bold italic underlined amino acids) and putative iron-responsive element (IRE; bold underlined) are indicated.
(b) Phylogenetic tree of selected members of the Dmt/Nramp family found in plants AtNramp1 (AF165125), AtNramp2 (AF141204), AtNramp3 (AF202539), AtNramp4 (AF202540), AtNRAMP5 (CAC27822), AtNramp6 (CAC28123), AtEin2 (AAD41076), OsNramp1 (S62667), OsNramp2 (AAB61961), OsNramp3 (AAC49720). The phylogenetic tree was drawn using MacVector (Accelrys) after comparison of deduced amino acid sequences using the CLUSTAL W method. The phylogram was built using the neighbour-joining method and best-tree mode. Distances between proteins were estimated using the Poisson-correction algorithm.
(c) Hydropathy analysis of the deduced amino acid sequence of GmDmt1;1 calculated using the Kyte and Doolittle algorithm with an amino acid window size of 19. Putative transmembrane spanning regions are indicated with horizontal bars. Dashed bar indicates hydrophilic section of protein used to generate anti-GmDmt1 antisera.

AY169405). Analysis of the GmDmt1;1 nucleotide sequence identified an open-reading frame of 516 amino acids encoding for a putative protein of approximately 57 kDa (Figure 1a). A BLAST search analysis of the GmDmt1;1 amino acid sequence identified significant homology (approximately 29% identity; approximately 46% similarity) to the amino acid sequences of six members of the *Arabidopsis* Nramp family (excluding AtEin2) of divalent metal ion transporters (Figure 1b). Hydropathy analysis (Kyte and Doolittle, 1982) of the encoded amino acids identified a protein with 12 putative transmembrane-spanning regions (Figure 1c). Between transmembrane segments 8 and 9, there is a conserved transport motif (5′-GQSSTITGTYAGQFIMGGFLN-3′), common among Nramp/Dmt homologues (Figure 1a). In the 3′-untranslated region of *GmDmt1;1*, there is an iron-responsive element (IRE) motif (5′-CTATGTCAGAG-3′) between bases 1688–1698 (Figure 1a).

A search of the Soybean TIGR Gene Index (http://www.tigr.org) yielded several soybean sequences similar to *GmDmt1;1*. These sequences consisted of expressed sequence tags (ESTs) aligned to make four tentative consensus sequences (TC84846, TC93163, TC94978 and TC82594), while a fifth sequence was from GenBank (accession no. AW277420). These partial sequences are between 65 and 98%, identical to *GmDmt1;1*. Sequence TC93163 has 98% identity with *GmDmt1;1* (isolated from cv. Stevens) and is likely to represent the same isoform from soybean cv. Williams. Obviously, *GmDmt1;1* is a member of a small gene family in soybean.

Gene expression

Northern blot analysis demonstrated that GmDmt1;1 is a nodule-enhanced protein. *GmDmt1;1* mRNA transcripts were abundant in nodules, but were only weakly detected in roots, leaves and stems (Figure 2a). Coincidently, nodule *GmDmt1;1* mRNA expression was the highest during the growth period, associated with maximum rates of symbiotic nitrogen fixation (20–40 days after planting), and decreased thereafter (Figure 2b,c). In young developing nodules, *GmDmt1;1* mRNA was barely detectable (Figure 2b).

Protein localisation

Antibodies were raised in rabbits against the N-terminal 73 amino acids of GmDmt1;1 (Figure 1c). This antiserum was used in Western blot analysis of 4-week-old total soluble nodule proteins, nodule microsomes, PBS proteins and PBM, isolated from purified symbiosomes. The anti-GmDMT1 antiserum identified a 67-kDa protein on the PBM-enriched nodule protein fraction (Figure 3a), but did not cross-react with soluble nodule proteins, PBS proteins or nodule microsomes (Figure 3a). Replicate Western blots incubated with pre-immune serum (Figure 3b) did not

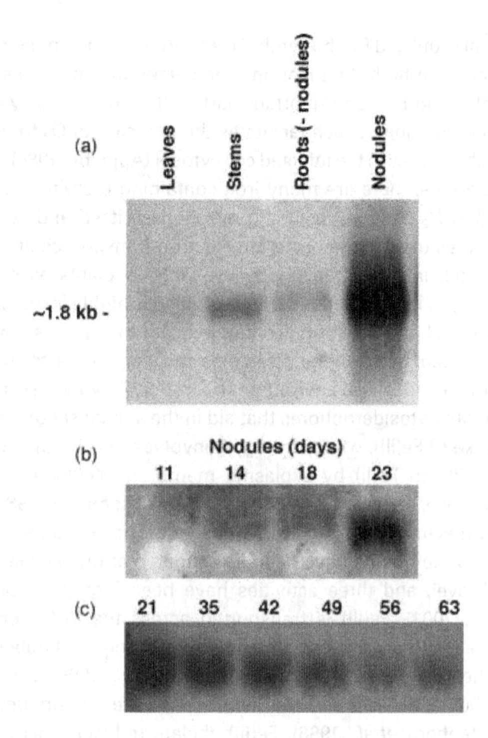

Figure 2. Northern blot analysis of *GmDmt1;1* expression.
(a) *GmDmt1;1* tissue expression. One microgram of poly(A)+-enriched RNA was extracted from 4-week-old soybean leaves, stems, roots (nodules detached) and nodules.
(b) GmDmt1;1 expression in developing nodules.
(c) GmDmt1;1 expression in mature nodules.
Ten micrograms of total RNA was extracted from the nodules prior to and after the onset of symbiotic nitrogen fixation. Blots (a) and (c) were probed with DIG-labelled antisense *GmDmt1;1* full-length RNA, while blot (b) was probed with randomly primed DIG-labelled full-length *GmDmt1;1* cDNA.

cross-react with the soybean nodule tissues examined. The protein identified on the PBM-enriched protein fraction is approximately 10 kDa larger than that predicted by the amino acid sequence of GmDmt1. The increase in size may be related to extensive post-translational modification (e.g. glycosylation) of GmDmt1, as it occurs in other systems. For example, the human Nramp1 and Nramp2 homologues are extensively modified by glycosylation and can appear about 40% larger on SDS–PAGE than predicted by their amino acid sequence alone (Gruenheid *et al.*, 1999; Tabuchi *et al.*, 2000, 2002). Post-translational modification of PBM proteins has been observed previously (Cheon *et al.*, 1994; Kaiser *et al.*, 1998), and the PBM protein Nod 24 undergoes extensive post-translation modification en route to the PBM, changing its apparent size on SDS–PAGE from 15 to 32 kDa (Cheon *et al.*, 1994). The localisation of GmDmt1;1 to the PBM was confirmed by subsequent immunogold-labelling experiments on fixed sections of infected cells containing symbiosomes. The anti-GmDmt1;1 antisera cross-reacted primarily with proteins on the PBM (Figure 3c,d).

Occasional cross-reactivity with bacteroids was also evident, but this was significantly reduced with more stringent blocking buffers, which included 5% w/v foetal albumin and 3% w/v normal goat serum (Figure 3e).

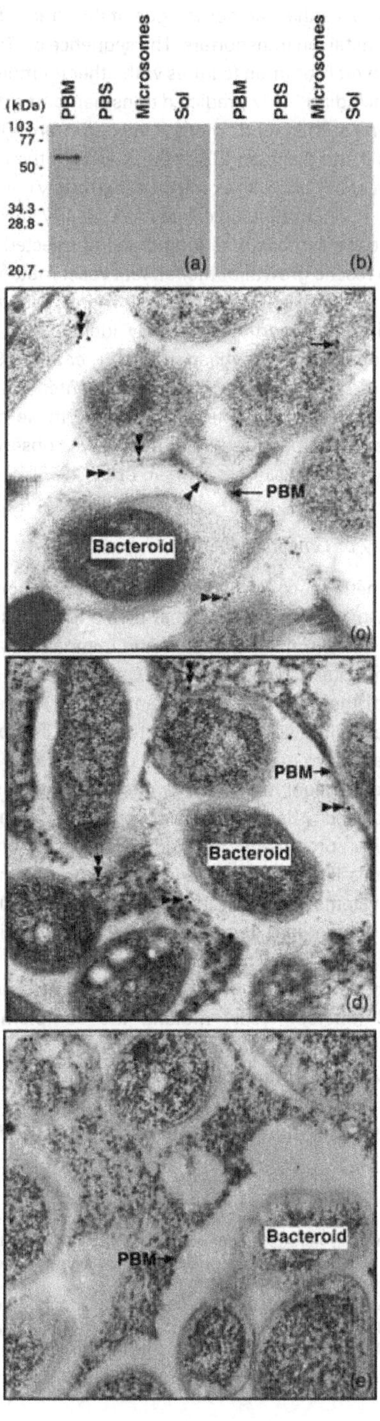

Figure 3. Immunolocalisation of GmDmt1;1 to the peribacteroid membrane (PBM) of soybean nodules.

Western analysis of SDS–PAGE separated and blotted 4-week-old nodule protein fractions including enriched PBM, peribacteroid space (PBS) proteins, total nodule microsomes and soluble proteins. Duplicate blots were incubated with anti-GmDmt1;1 antiserum (a) or with pre-immune antisera (b) at a dilution of 1 : 3000, respectively. Thirty micrograms of purified protein was loaded in each lane. Molecular size markers are shown on the left. (c–e) Immunogold labelling of 3-week-old soybean nodule cross-sections of infected cells with symbiosomes. Tissue sections were incubated with anti-GmDmt1 antisera at a dilution of 1 : 100 (c, d) or with the pre-immune serum at a dilution of 1 : 50 (e) followed by 15-nm colloidal gold conjugated with goat antirabbit IgG (BIOCELL EM GAR 15) at a dilution of 1 : 40. Double arrows indicate immunoreactive proteins on the PBM and single arrows identify possible cross-contamination with bacteroids. EM magnification for both pictures was 35 000×.

Functional analysis in yeast

To test for Fe^{2+}-transport activity, GmDmt1;1 and the positive control AtIrt1 (a known iron transporter) was cloned into the yeast-expression vectors, pFL61 and pDR195, and then transformed into the yeast iron-transport mutant DEY1453 (*fet3fet4*), which grows poorly on media containing low iron concentrations as a result of disrupted high (*fet3*)- and low (*fet4*)-affinity Fe^{2+}-transport activity (Dix *et al.*, 1994; Eide *et al.*, 1992). On synthetic-defined (SD) media supplemented with or without 2 μM $FeCl_3$, both AtIrt1 and GmDmt1;1 improved the growth of *fet3fet4* cells over those containing the empty cloning vector pFL61 (Figure 4a). Similarly, in liquid SD media supplemented with 20 μM $FeCl_3$ cells containing either AtIrt1 or GmDmt1;1 routinely entered the exponential-growth phase earlier than those of the empty vector controls (Figure 4b). In the absence of any added iron, GmDmt1;1 was unable to enhance growth of the mutant yeast (results not shown).

Short-term uptake experiments with 1 μM $^{55}FeCl_3$ showed that transformation of *fet3fet4* cells with GmDmt1;1 enhanced accumulation of ^{55}Fe(II) approximately fourfold over control cells (Figure 5a). This uptake followed Michaelis–Menten kinetics with an apparent K_M of 6.4 ± 1.1 μM (Figure 5b). The apparent K_M for Fe(II) agrees well with the need for supplementation of growth medium with micromolar iron in order to observe enhanced growth by the GmDmt1;1 cells (see above).

We tested whether GmDmt1;1 can transport other metal ions by heterologous expression in the zinc-deficient yeast-transport mutant, ZHY3 (*zrt1zrt2*) and the manganese transport mutant SMF1 (Chen *et al.*, 1999). On minimal zinc plates, GmDmt1 partially complemented ZHY3, but the growth of this mutant was slower than that of DEY1453 (*fet3fet4*) transformed with GmDmt1;1 (mean doubling times were 6.3 ± 0.5 h versus 5.1 ± 0.01 h ($n = 4$), respectively). In short-term transport studies, a 10-fold excess of $MnCl_2$ in the reaction medium inhibited ^{55}Fe uptake significantly by DEY1453 (*fet3fet4*) transformed with GmDmt1;1 (Figure 5c). Similar inhibitions were seen with 10-fold $CuCl_2$ and $ZnCl_2$ (Figure 5c).

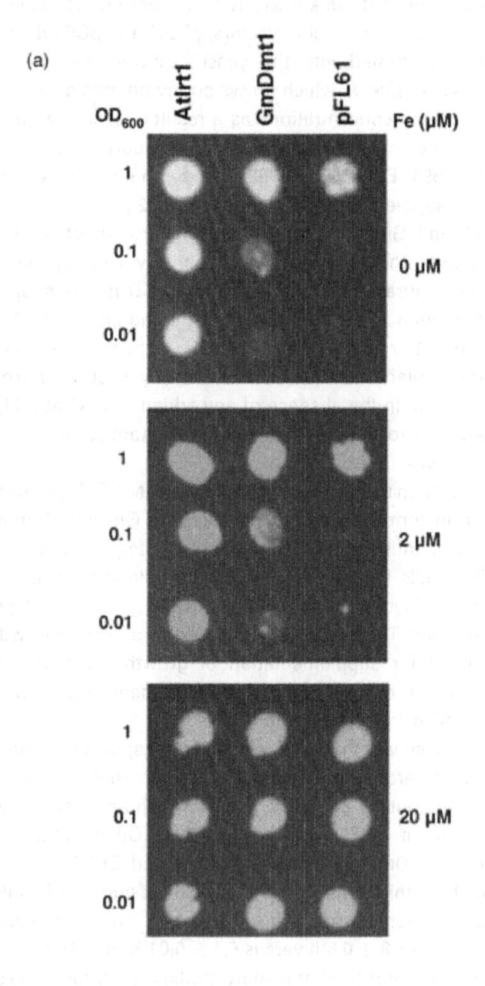

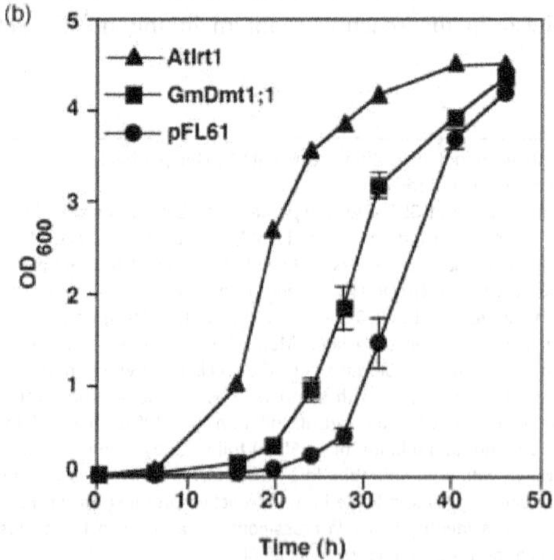

Figure 4. Functional analysis of GmDmt1;1 activity in yeast cells.
fet3fet4 yeast cells were transformed with GmDmt1;1 inserted in the expression vector pFL61. Cells were also transformed with empty yeast expression vectors.
(a) Growth of serially diluted cells after 6 days at 30°C of GmDmt1;1 (GmDmt1;1-pFL61), Atlrt1 (Atlrt1-pFL61) and control (pFL61) transformed *fet3fet4* cells on synthetic-defined (SD) media supplemented with 0, 2, 20 μM $FeCl_3$.
(b) Growth in liquid SD media supplemented with 20 μM $FeCl_3$.

Discussion

GmDmt1;1 can transport ferrous iron

The results presented here demonstrate that GmDmt1;1 is a symbiotically enhanced homologue of the Nramp family of divalent metal ion transporters. The sequence of *GmDmt1;1* shares several common features with other members of the family, including 11–12 predicted transmembrane domains, a consensus transport motif between transmembrane domains 8 and 9 and an IRE in the 3′-UTR of the transcript (Gunshin *et al.*, 1997). Its expression is strongly enhanced in nodules, and immunological studies clearly localise the protein to the symbiosome membrane of infected cells. Its ability to rescue growth of the *fet3fet4* yeast mutant on low iron medium makes GmDmt1;1 a strong candidate for the ferrous iron transporter, previously identified in isolated symbiosomes from soybean (Moreau *et al.*, 1998). The kinetics of $^{55}Fe^{2+}$ uptake into complemented yeast (with an apparent K_M of 6.4 μM) also resemble those observed in isolated symbiosomes (linear uptake was observed over the range of 5–50 μM iron; Moreau *et al.*, 1998).

Specificity of GmDmt1;1

The competition experiments shown in Figure 5(c) indicate that GmDmt1 can transport other divalent cations in addition to ferrous iron. Zinc, copper and manganese all inhibited iron uptake. The ability of GmDmt1;1 to enhance growth of the *zrt1zrt2* yeast mutant further suggests that the protein is not specific for iron transport. The preferred substrate *in vivo* may well depend on the relative concentrations of divalent metals in the infected cell cytosol. This lack of specificity has been found with Nramp homologues from other organisms, including Nramp2 from mice. Despite this lack of specificity when expressed in heterologous systems, mutation of murine Nramp2 results in an anaemic phenotype, demonstrating that *in vivo* it is predominantly an iron transporter (Fleming *et al.*, 1997). Although GmDmt1;1 was able to complement the DEY1453 (*fet3fet4*) yeast mutant, the complementation was not robust and the growth media had to be supplemented with low concentrations of iron. Atlrt1, on the other hand, showed much better complementation and allowed growth of the mutant in the absence of added iron

(Figure 4). There are several possible reasons for the poorer growth with GmDmt1;1, including possible instability of GmDmt1;1 transcripts (perhaps because of the presence of the regulatory IRE element in the transcript).

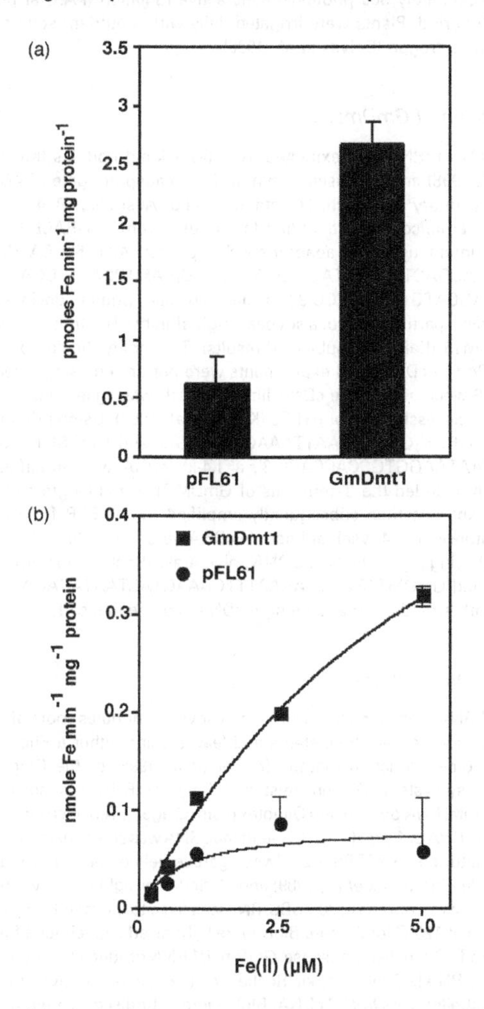

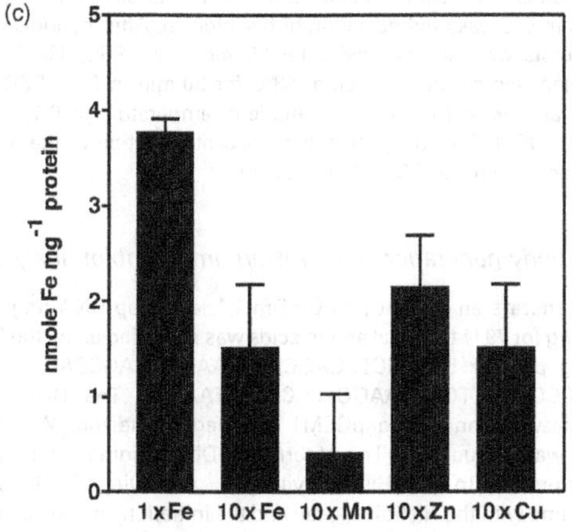

Localisation and function of GmDmt1;1

It has been suggested that AtNramp has an intracellular localisation (Grotz and Guerinot, 2002). The symbiosome is a vacuole-like structure (Mellor, 1989) and contains high concentrations of non-heme iron (Wittenberg *et al.*, 1996). However, this raises an interesting question as to the mechanism of GmDmt1;1. Divalent metal transport into vacuoles is likely to occur as Fe^{2+}/H^+ exchange (Gonzalez *et al.*, 1999), and it is possible that this also occurs in symbiosomes, as the PBM is energised by a H^+-pumping ATPase, which generates a membrane potential positive on the inside (and an acidic interior if permeant anions are present; Udvardi and Day, 1997). However, in this situation, and also in yeast, GmDmt1;1 catalyses uptake of iron into the cell, while uptake into symbiosomes is equivalent to export from the plant cytosol. Assuming that GmDmt1;1 is located in the plasma membrane of yeast and that it has the same physical orientation as in symbiosomes, which is likely considering that the secretory pathway is thought to mediate protein insertion into the PBM, then GmDmt1;1 must be able to catalyse bidirectional transport of iron. This is not unusual for a carrier and has been observed with GmZip1, a zinc transporter on the PBM. It appears that iron uptake can be linked to the membrane potential or pH gradient via other ion movements in the heterologous system. Further experiments on symbiosomes and yeast (or *Xenopus* oocytes) may provide new insights into the mechanism of iron transport in plants, but it appears that GmDmt1;1 has the capacity to function *in vivo* as either an uptake or an efflux mechanism in symbiosomes. This also raises the question of the relationship between GmDmt1;1 and the NADH-ferric chelate reductase on the PBM (Levier and Guerinot, 1996).

At the plant plasma membrane, ferrous iron transporters (presumably AtIrt1 homologues) act to take up iron reduced by the reductase into the plant. In the symbiosome, assuming that the orientation of the reductase on the PBM is similar to that on the plasma membrane, ferric iron stored in the symbiosome space would be reduced upon oxidation

Figure 5. Uptake of Fe(II) by GmDmt1 in yeast.
(a) Influx of $^{55}Fe^{2+}$ into yeast cells transformed with GmDmt1;1. *fet3fet4* cells were transformed with GmDmt1;1-pFL61 or pFL61 and then incubated with 1 μM $^{55}FeCl_3$ (pH 5.5) for 5- and 10-min periods. Data presented are means ± SE of ^{55}Fe uptake between 5 and 10 min from three separate experiments (each performed in triplicate).
(b) Concentration dependence of ^{55}Fe influx into *fet3fet4* cells transformed with GmDmt1;1-pFL61 or pFL61. Data presented are means ± SE of ^{55}Fe uptake over 5 min ($n = 3$). The curve was obtained by direct fit to the Michaelis–Menten equation. Estimated K_M and V_{MAX} for GmDmt1;1 were 6.4 ± 1.1 μM Fe(II) and 0.72 ± 0.08 nM Fe(II) min^{-1} mg^{-1} protein, respectively.
(c) Effect of other divalent cations on uptake of $^{55}Fe^{2+}$ into *fet3fet4* cells transformed with pFL61-*Gm*DMT1;1. Data presented are means ± SE of ^{55}Fe (10 μM) uptake over 10 min in the presence and absence of 100 μM unlabelled Fe^{2+}, Cu^{2+}, Zn^{2+} and Mn^{2+}.

of NADH in the plant cytosol. In isolated symbiosomes, addition of NADH together with ferric citrate, stimulated iron accumulation in the bacteroid, suggesting that the ferrous iron produced in the symbiosome space was taken up by the bacteroid ferrous iron transporter (Moreau *et al.*, 1998). *In vivo*, however, Fe(II) in the symbiosome space could also be transported back into the plant cytosol by the action of GmDmt1;1. We attempted to demonstrate this with isolated symbiosomes by loading them with $^{55}Fe^{3+}$ citrate, adding NADH and ATP (the latter to energise the membrane), and measuring efflux of ^{55}Fe into the reaction medium, but could not detect any efflux (Thomson, data not shown). The direction of transport *in vivo* will depend on the concentration of other ions on either side of the PBM and the activity of the bacteroid ferric and ferrous transporters.

Regulation of GmDmt1;1 expression

As mentioned above, GmDmt1;1 contains an IRE in its 3′-UTR. IREs are conserved sequences in the UTR of certain RNA transcripts to which iron-regulating proteins (IRPs) bind. The presence of an IRE motif suggests that *GmDmt1;1* mRNA may be stabilised by the binding of IRPs in soybean nodules when free iron levels are low. In both mammals (Canonne-Hergaux *et al.*, 1999) and *Årabidopsis* (Curie *et al.*, 2000; Thomine *et al.*, 2000), the abundance of Dmt isoforms containing an IRE element is enhanced by iron deficiency. Iron is required for both plant and bacterial enzymes during nodule development and in the functioning of the mature nodule. *GmDmt1;1* transcripts were detectable in relatively young (11-day-old) nodules and increased as the nodules matured (Figure 2). It is possible that during this time, when the bacteroid and plant iron requirements are relatively high, free iron levels are low and *GmDMT1* transcripts are stabilised by IRPs. This process could ensure nodule iron transport capacity through increased expression and activity of GmDMT1.

Conclusion

We have identified an Nramp homologue, GmDmt1, which is expressed in soybean nodules and encodes a divalent metal ion transporter located on the symbiosome membrane. The ability of this protein to transport ferrous iron makes it a candidate for the ferrous transport activity previously demonstrated in isolated symbiosomes (Moreau *et al.*, 1998).

Experimental procedures

Plant growth

Soybean (*Glycine max* L. cv. Stevens) seeds were inoculated at planting with *Bradyrhizobium japonicum* USDA 110 and grown in river sand in either glass houses under ambient light between 20 and 30°C, or in controlled-temperature growth rooms at 25°C day and 21°C night temperatures. Plants in the growth chambers were provided with a scheduled (14-h day/10-h night) artificial light (approximately 300 photosynthetic active radiation (PAR) at pot level) period. Plants were irrigated daily with a nutrient solution lacking nitrogen (Delves *et al.*, 1986).

Isolation of GmDmt1;1

Poly(A)+ mRNA was extracted from 6-week-old nodules (Kaiser *et al.*, 1998) and was used to synthesise an adaptor-ligated RACE cDNA library (Clontech; Marathon, Roche, Australia). A 480-bp cDNA amplicon was identified fortuitously from a 5′-RACE PCR experiment using an adaptor-specific primer, AP1: 5′-CCATCC-TAATACGACTCACTATAGGGC-3′ and GmAMTR24: 5′-CGAAC-CAAAGCATGAAGGTCCC-3′, a gene-specific primer designed against a partial cDNA of a soybean high-affinity NH_4^+ transporter, GmAMT1 (Kaiser, unpublished results). To amplify the complete *GmDmt1;1* cDNA, PCR experiments were performed using a second 6-week-old nodule cDNA library, which was ligated into the yeast-expression vector pYES3 (Kaiser *et al.*, 1998). Using primers pYES11R: 5′-GCCGCAAATTAAAGCCTTCG-3′ and GmDMTF2: 5′-AAGAATAAGGTGCCACCACC-3′, a 1.4-kb cDNA was amplified, which included the 3′-terminus of GmDMT1. A full-length clone (1.88 kb) was then subsequently amplified by the PCR from an adaptor-ligated 4-week-old nodule cDNA library (Clontech; Marathon) using high-fidelity Taq DNA polymerase (Roche) and primers AP1 and GmDMT1R21: 5′-AAAATTTGAAAGTACTAATACAGAGC-3′. Both strands of the full-length cDNA were sequenced.

Northern analysis

Total RNA was extracted from frozen soybean nodules roots after nodules were detached, stems and leaves using either a Phenol/Guanidine extraction method (Kaiser *et al.*, 1998) or the Qiagen RNAeasy system (Qiagen, Australia). Poly(A)+ RNA was isolated from total RNA pools using Oligotex resin (Qiagen). Ten micrograms of total RNA or 1 μg of Poly(A)+-enriched RNA was size-separated on a denaturing 1X MOPS 1.2% (w/v) agarose gel containing formaldehyde (Sambrook *et al.*, 1989) and blotted overnight onto Hybond N^+ nylon membrane in 20× SSC. RNA was fixed to the membrane by baking at 120°C for 30 min. Blots were hybridised with either a full-length DIG-labelled antisense *GmDmt1;1* RNA produced using the SP6/T7 RNA DIG-labelling kit (Roche) or full-length randomly primed DIG-labelled *GmDmt1;1* cDNA. Blots were hybridised overnight at 68°C in DIG-easy hybridisation buffer (Roche). After hybridisation, the blots were washed twice for 15 min in 2× SSC, 1% SDS at ambient temperature, twice at 68°C for 30 min in 0.1× SSC, 1% SDS and twice for 15 min at ambient temperature in 0.1× SSC, 0.1% SDS, followed by chemiluminescent detection of the digoxygenin label using CDP-STAR (Roche).

Antibody generation and Western immunoblot analysis

To generate an antibody to GmDmt1;1, a 236-bp DNA fragment coding for 79 N-terminal amino acids was amplified using the PCR, using primers 5′-TGGCTCGAGCCACCAAGAGCAGCCACT-3′ and 5′-ACCCGAATTCCTGAAGGTCCCCCTCTAAG-3′. The DNA fragment was cloned into pGEMT (Promega, Madison, WI, USA) and was sequenced. The N-terminal DNA fragment was then subcloned into pTrcHisB (Invitrogen, San Diego, CA, USA) in-frame with the $Histidine_{(6)}$-tag and the initiation and termination codon. The resulting construct, pHISDMT1, was transformed into *Escherichia coli* TOP10F′ cells (Invitrogen) and grown in 500 ml of liquid Solution B (SOB) media containing 50 μg ml^{-1} ampicillin at 37°C to an OD_{600} of 0.5. Expression of the $His_{(6)}$-tag GmDmt1;1 fusion protein was then induced by adding 1 mM isopropyl β-D-thiogalactopyranoside (IPTG) and incubating further for 3 h. Cells were collected and lysed in buffer (8 M urea, 50 mM NaH_2PO_4, 300 mM NaCl, 1.5 mM imidazole pH 8.0) and disrupted by six cycles of freezing and thawing followed by repeated passage through an 18-gauge needle. Insoluble proteins and cell debris were removed by centrifugation for 10 min at 16 000 ***g***, and the supernatant was collected. The $His_{(6)}$-tagged GmDmt1;1 fusion protein was purified by immobilised metal affinity chromatography (Clontech, San Diego, CA, USA). Eluted protein was concentrated by tricholoracetic acid precipitation and re-suspended in elution buffer containing 8 M urea. The concentrated fusion protein (approximately 200 μg) was mixed with an equal volume of complete Freunds adjuvant (Sigma, USA) and injected into New Zealand White rabbits followed by four subsequent 200-μg injections at 1-month intervals. Ten days after the final injection, crude serum was collected. Protein fractions for Western immunoblot analysis were separated by 12 or 15% w/v SDS–PAGE (Laemmli, 1970) and blotted onto Polyvinylidene Fluoride (PVDF) membranes (Amersham, Buckinghamshire, UK), using a wet-blotting system (Bio-Rad, Regents Park, Australia). Membranes were probed with antiserum to GmDmt1;1 at a dilution of 1 : 3000 in PBS buffer, followed by secondary probing with a horseradish peroxidase-conjugated antirabbit IgG antibody. Immunoreactive proteins were visualised by chemiluminescence using a commercial kit (Roche, Australia).

Symbiosome isolation and nodule membrane purification

Symbiosomes were purified from soybean nodule extracts as described before (Day *et al.*, 1989), using a 3-step Percoll gradient. PBM-enriched membrane fractions were purified by rapid vortexing (4 min) of symbiosomes in buffer (350 mM mannitol, 25 mM MES-KOH (pH 7.0), 3 mM $MgSO_4$, 1 mM PMSF; 1 mM pAB; 10 μM E64; 1 mM DTT), followed by centrifugation at 10 000 ***g*** for 10 min in a SS34 rotor (4°C). The supernatant was collected and centrifuged further at 125 000 ***g*** for 60 min to separate the PBS proteins from the insoluble PBM-enriched membrane fraction. The PBM pellet was phenol-extracted (Hurkman and Tanaka, 1986), and the PBM and PBS fractions were concentrated by ammonium acetate/methanol precipitation and re-suspended at room temperature in loading buffer (125 mM Tris pH 6.8, 4% w/v SDS, 20% v/v glycerol, 50 mM DTT, 20% v/v mercaptoethanol, 0.001% w/v bromophenol blue). Soluble and insoluble nodule fractions were prepared by grinding nodules in buffer (25 mM MES-KOH pH 7.0, 350 mM mannitol, 3 mM $MgSO_4$, 1 mM PMSF, 1 mM pAB; 10 μM E64), followed by filtration through four layers of miracloth (Calbiochem, San Diego, CA, USA), and were centrifuged at 10 000 ***g***, 4°C for 15 min to separate the bacteroids from the plant fraction. The supernatant was centrifuged further at 125 000 ***g***, 4°C for 1 h. The supernatant was collected and concentrated by ammonium acetate/methanol precipitation. The nodule total membrane pellet and soluble protein fractions were re-suspended in loading buffer as described above.

Functional expression in yeast

GmDmt1;1 was cloned into the *NotI* site of the yeast–*E. coli* shuttle vector pDR195 downstream of the P-type ATPase promoter PMA1 (Thomine *et al.*, 2000) or into pFL61 under the control of the phosphoglycerate kinase promoter (Minet *et al.*, 1992). Yeast strain DEY1453 (*fet3fet4*) (Eide *et al.*, 1996) (*MATa/MATα ade2/+can1/can1 his3/his3 leu2/leu2 trp1/trp1 ura3/ura3 fet3-2::HIS3/fet3-2::HIS3/fet4-1::LEU2/fet4-1::LEU2*) was transformed (Gietz *et al.*, 1992) and selected for growth on SD media containing 20 mg ml^{-1} glucose and appropriate autotrophic requirements (pH 4.5; Dubois and Grenson, 1979). The media was also supplemented with 10 μM $FeCl_3$ to aid in the growth of *fet3fet4*. Yeast-uptake experiments were performed based on the protocol of Eide *et al.* (1992). fet3fet4 cells transformed with expression plasmids were grown to log phase in SD media with 2 μM additional $FeCl_3$. Log-phase cells were harvested, washed in H_2O and diluted in new SD media to an OD_{600} of 0.3 and grown for a further 4 h. Cells were harvested and washed twice with cold MES Glucose Nitriso-acetic acid (MGN) uptake buffer (10 mM MES, pH 5.5, 2% (w/v) glucose, 1 mM nitrilotriacetic acid). Cells were equilibrated at 30°C for 10 min before addition of an equal volume of $^{55}Fe^{2+}$ solution (MGN buffer, with 10 μM $FeCl_3$, $^{55}FeCl_3$ and 200 μM ascorbic acid to ensure that iron is in the ferrous form). Cells were incubated at 30°C, and aliquots were taken, filtered and washed five times with 500-μl ice-cold synthetic seawater medium (SSW) (1 mM EDTA, 20 mM trisodium citrate, 1 mM KH_2PO_4, 1 mM $CaCl_2$, 5 mM $MgSO_4$, 1 mM NaCl (pH 4.2)). Duplicate experiments were performed on ice as a background control for iron binding to cellular material. Internalised $^{55}Fe^{2+}$ was determined by liquid scintillation counting of the filters. Protein amounts were determined using a modified Lowry assay (Peterson, 1977).

Acknowledgements

This research was financially supported by a grant from the Australian Research Council (D.A. Day), the CNRS Programme International de Cooperation Scientifique, Program 637 (S. Moreau, A. Puppo) and a Canadian National Science and Engineering Research Council Postdoctoral fellowship (B.N. Kaiser). We thank Ghislaine Van de Sype for expert technical assistance with the microscopy.

References

Alonso, J.M., Hirayama, T., Roman, G., Nourizadeh, S. and Ecker, J.R. (1999) EIN2, a bifunctional transducer of ethylene and stress responses in *Arabidopsis. Science*, **284**, 2148–2152.

Appleby, C.A. (1984) Leghemoglobin and rhizobium respiration. *Annu. Rev. Plant Physiol.* **35**, 443–478.

Belouchi, A., Kwan, T. and Gros, P. (1997) Cloning and characterization of the OsNramp family from *Oryza sativa*, a new family of membrane proteins possibly implicated in the transport of metal ions. *Plant Mol. Biol.* **33**, 1085–1092.

Canonne-Hergaux, F., Gruenheid, S., Govoni, G. and Gros, P. (1999) The Nramp1 protein and its role in resistance to infection and macrophage function. *Proc. Assoc. Am. Physicians*, **111**, 283–289.

Chen, X.Z., Peng, J.B., Cohen, A., Nelson, H., Nelson, N. and Hediger, M.A. (1999) Yeast SMF1 mediates H^+-coupled iron uptake with concomitant uncoupled cation currents. *J. Biol. Chem.* **274**, 35089–35094.

Cheon, C., Hong, Z. and Verma, D.P.S. (1994) Nodulin-24 follows a novel pathway for integration into the peribacteroid membrane in soybean root nodules. *J. Biol. Chem.* **269** (9), 6598–6602.

Curie, C., Alonso, J.M., Le Jean, M., Ecker, J.R. and Briat, J.F. (2000) Involvement of NRAMP1 from *Arabidopsis thaliana* in iron transport. *Biochem. J.* **347**, 749–755.

Day, D.A., Price, G.D. and Udvardi, M.K. (1989) Membrane interface of the *Bradyrhizobium japonicum–Glycine max* symbiosis: peribacteroid units from soybean nodules. *Aust. J. Plant Physiol.* **16**, 69–84.

Day, D.A., Kaiser, B.N., Thomson, R., Udvardi, M.K., Moreau, S. and Puppo, A. (2001) Nutrient transport across symbiotic membranes from legume nodules. *Aust. J. Plant Physiol.* **28**, 667–674.

Delves, A.C., Matthews, A., Day, D.A., Carter, A.S., Carroll, B.J. and Gresshoff, P.M. (1986) Regulation of the soybean – rhizobium nodule symbiosis by shoot and root factors. *Plant Physiol.* **82**, 588–590.

Dix, D.R., Bridgham, J.T., Broderius, M.A., Byersdorfer, C.A. and Eide, D.J. (1994) The *fet4* gene encodes the low-affinity Fe(II) transport protein of *Saccharomyces cerevisiae*. *J. Biol. Chem.* **269**, 26092–26099.

Dubois, E. and Grenson, M. (1979) Methylamine/ammonium uptake systems in *Saccharomyces cerevisiae*: multiplicity and regulation. *Mol. Gen. Genet.* **175**, 67–76.

Eide, D., Davis-Kaplan, S., Jordan, I., Sipe, D. and Kaplan, J. (1992) Regulation of iron uptake in *Saccharomyces cerevisiae*. *J. Biol. Chem.* **267**, 20774–20781.

Eide, D., Broderius, M., Fett, J. and Guerinot, M.L. (1996) A novel iron-regulated metal transporter from plants identified by functional expression in yeast. *Proc. Natl. Acad. Sci. USA*, **93**, 5624–5628.

Fleming, M.D., Trenor, C.C., Su, M.A., Foernzler, D., Beier, D.R., Dietrich, W.F. and Andrews, N.C. (1997) Microcytic anaemia mice have a mutation in *Nramp2*, a candidate iron transporter gene. *Nat. Genet.* **16**, 383–386.

Gietz, D., StJean, A., Woods, R.A. and Schiestl, R.H. (1992) Improved method for high-efficiency transformation of intact yeast cells. *Nucl. Acids Res.* **20**, 1425–1420.

Gonzalez, A., Koren'kov, V. and Wagner, G.J. (1999) A comparison of Zn, Mn, Cd and Ca-transport mechanisms in oat root tonoplast vesicles. *Physiologia Plantarum*, **106**, 203–209.

Grotz, N. and Guerinot, M.L. (2002) Limiting nutrients: an old problem with new solutions? *Curr. Opin. Plant Biol.* **5**, 158–163.

Gruenheid, S., Canonne-Hergaux, F., Gauthier, S., Hackam, D.J., Grinstein, S. and Gros, P. (1999) The iron-transport protein NRAMP2 is an integral membrane glycoprotein that colocalizes with transferrin in recycling endosomes. *J. Exp. Med.* **189**, 831–841.

Gunshin, H., Mackenzie, B., Berger, U.V., Gunshin, Y., Romero, M.F., Boron, W.F., Nussberger, S., Gollan, J.L. and Hediger, M.A. (1997) Cloning and characterization of a mammalian proton-coupled metal ion transporter. *Nature*, **388**, 482–488.

Hurkman, W.J. and Tanaka, C.K. (1986) Solubilization of plant-membrane proteins for analysis by two-dimensional gel electrophoresis. *Plant Physiol.* **81**, 802–806.

Kaiser, B.N., Finnegan, P.M., Tyerman, S.D., Whitehead, L.F., Bergersen, F.J., Day, D.A. and Udvardi, M.K. (1998) Characterization of an ammonium transport protein from the peribacteroid membrane of soybean nodules. *Science*, **281**, 1202–1206.

Kyte, J. and Doolittle, R.F. (1982) A simple method for displaying the hydropathic character of a protein. *J. Mol. Biol.* **157**, 105–132.

Laemmli, U.K. (1970) Cleavage of structural proteins during the assembly of the head of bacteriophage T4. *Nature*, **227**, 680–685.

Levier, K., Day, D.A. and Guerinot, M.L. (1996) Iron uptake by symbiosomes from soybean root nodules. *Plant Physiol.* **111**, 893–900.

Levier, K. and Guerinot, M.L. (1996) The *Bradyrhizobium japonicum fega* gene encodes an iron-regulated outer membrane protein with similarity to hydroxamate-type siderophore receptors. *J. Bacteriol.* **178**, 7265–7275.

Mellor, R.B. (1989) Bacteroids in the *Rhizobium*-legume symbiosis inhabit a plant internal lytic compartment: implications for other microbial endosymbioses. *J. Exp. Bot.* **40**, 831–839.

Minet, M., Dufour, M. and Lacroute, F. (1992) Complementation of *Saccharomyces cerevisiae* auxotrophix mutants by *Arabidopsis thaliana* cDNAs. *Plant J.* **2**, 417–422.

Moreau, S., Meyer, J.M. and Puppo, A. (1995) Uptake of iron by symbiosomes and bacteroids from soybean nodules. *FEBS Lett.* **361**, 225–228.

Moreau, S., Day, D.A. and Puppo, A. (1998) Ferrous iron is transported across the peribacteroid membrane of soybean nodules. *Planta*, **207**, 83–87.

Moreau, S., Thomson, R.M., Kaiser, B.N., Trevaskis, B., Guerinot, M.L., Udvardi, M.K., Puppo, A. and Day, D.A. (2002) GmZIP1 encodes a symbiosis-specific zinc transporter in soybean. *J. Biol. Chem.* **277**, 4738–4746.

Peterson, G.L. (1977) A simplification of the protein assay of Lowry et al. which is more generally applicable. *Anal. Biochem.* **83**, 346–356.

Rodrigues, V., Cheah, P.Y., Ray, K. and Chia, W. (1995) Malvolio, the *Drosophila* homologue of mouse Nramp-1 (bcg), is expressed in macrophages and in the nervous system and is required for normal taste behaviour. *EMBO J.* **14**, 3007–3020.

Romheld, V. (1987) Different strategies for iron acquisition in higher plants. *Physiol. Plant.* **70**, 231–234.

Sambrook, J., Fritsch, E.F. and Maniatis, T. (1989) *Molecular Cloning: a Laboratory Manual.* Cold Spring Harbour; Cold Spring Harbour Laboratory Press.

Supek, F., Supekova, L., Nelson, H. and Nelson, N. (1996) A yeast manganese transporter related to the macrophage protein involved in conferring resistance to mycobacteria. *Proc. Natl. Acad. Sci. USA*, **93**, 5105–5110.

Tabuchi, M., Yoshimori, T., Yamaguchi, K., Yoshida, T. and Kishi, F. (2000) Human NRAMP/DMT1, which mediates iron transport across endosomal membranes, is localised to late endosomes and lysosomes in HEp-2 cells. *J. Biol. Chem.* **275**, 22220–22228.

Tabuchi, M., Tanaka, N., Nishida-Kitayama, J., Ohno, H. and Kishi, F. (2002) Alternative splicing regulates the subcellular localisation of divalent metal transporter 1 isoforms. *Mol. Biol. Cell*, **13**, 4371–4387.

Tang, C., Robson, A.D. and Dilworth, M.J. (1990) A split-root experiment shows that iron is required for nodule initiation in *Lupinus angustifolius* L. *New Phytol.* **115**, 61–67.

Thomine, S., Wang, R.C., Ward, J.M., Crawford, N.M. and Schroeder, J.I. (2000) Cadmium and iron transport by members of a plant metal transporter family in *Arabidopsis* with homology to Nramp genes. *Proc. Natl. Acad. Sci. USA*, **97**, 4991–4996.

Udvardi, M.K. and Day, D.A. (1997) Metabolite transport across symbiotic membranes of legume nodules. *Annu. Rev. Plant Physiol. Plant Mol. Biol.* **48**, 493–523.

Vert, G., Grotz, N., Dedaldechamp, F., Gaymard, F., Guerinot, M.L., Briata, J.F. and Curie, C. (2002) IRT1, an *Arabidopsis* transporter essential for iron uptake from the soil and for plant growth. *Plant Cell*, **14**, 1223–1233.

Vidal, S.M. and Gros, P. (1994) Resistance to infection with intracellular parasites – identification of a candidate gene. *News Physiol. Sci.* **9**, 178–183.

Whitehead, L.F. and Day, D.A. (1997) The peribacteroid membrane. *Physiologia Plantarum*, **100**, 30–44.

Wittenberg, J.B., Wittenberg, B.A., Day, D.A., Udvardi, M.K. and Appleby, C.A. (1996) Siderophore-bound iron in the peribacteroid space of soybean root nodules. *Plant Soil*, **178**, 161–169.

CHAPTER 20

PEA2：Britton-Simmons and Abbott (2008)

Journal of Ecology 2008, **96**, 68–77 doi: 10.1111/j.1365-2745.2007.01319.x

Short- and long-term effects of disturbance and propagule pressure on a biological invasion

Kevin H. Britton-Simmons* and Karen C. Abbott†

Department of Ecology and Evolution, The University of Chicago, 1101 East 57th Street, Chicago, IL 60637, USA

Summary

1. Invading species typically need to overcome multiple limiting factors simultaneously in order to become established, and understanding how such factors interact to regulate the invasion process remains a major challenge in ecology.

2. We used the invasion of marine algal communities by the seaweed *Sargassum muticum* as a study system to experimentally investigate the independent and interactive effects of disturbance and propagule pressure in the short term. Based on our experimental results, we parameterized an integrodifference equation model, which we used to examine how disturbances created by different benthic herbivores influence the longer term invasion success of *S. muticum*.

3. Our experimental results demonstrate that in this system neither disturbance nor propagule input alone was sufficient to maximize invasion success. Rather, the interaction between these processes was critical for understanding how the *S. muticum* invasion is regulated in the short term.

4. The model showed that both the size and spatial arrangement of herbivore disturbances had a major impact on how disturbance facilitated the invasion, by jointly determining how much space-limitation was alleviated and how readily disturbed areas could be reached by dispersing propagules.

5. *Synthesis.* Both the short-term experiment and the long-term model show that *S. muticum* invasion success is co-regulated by disturbance and propagule pressure. Our results underscore the importance of considering interactive effects when making predictions about invasion success.

Key-words: biological invasion, biotic resistance, disturbance, establishment probability, propagule pressure, *Sargassum muticum*

Introduction

Biological invasions are a global problem with substantial economic (Pimentel *et al.* 2005) and ecological (Mack *et al.* 2000) costs. Research on invasions has provided important insights into the establishment, spread and impact of non-native species. One key goal of invasion biology has been to identify the factors that determine whether an invasion will be successful (Williamson 1996). Accordingly, ecologists have identified several individual factors (e.g. disturbance and propagule pressure) that appear to exert strong controlling influences on the invasion process. However, understanding how these processes interact to regulate invasions remains a major challenge in ecology (D'Antonio *et al.* 2001; Lockwood *et al.* 2005; Von Holle & Simberloff 2005).

Propagule pressure is widely recognized as an important factor that influences invasion success (MacDonald *et al.* 1989; Simberloff 1989; Williamson 1996; Lonsdale 1999; Cassey *et al.* 2005). Previous studies suggest that the probability of a successful invasion increases with the number of propagules released (Panetta & Randall 1994; Williamson 1989; Grevstad 1999), with the number of introduction attempts (Veltman *et al.* 1996), with introduction rate (Drake *et al.* 2005), and with proximity to existing populations of invaders (Bossenbroek *et al.* 2001). Moreover, propagule pressure may influence invasion dynamics after establishment by affecting the capacity of non-native species to adapt to their new environment (Ahlroth *et al.* 2003; Travis *et al.* 2005). Despite its acknowledged importance, propagule pressure has rarely been manipulated experimentally and the interaction of propagule pressure with other processes that regulate invasion success is not well understood (D'Antonio *et al.* 2001; Lockwood *et al.* 2005).

*Correspondence and present address. Friday Harbor Laboratories, University of Washington, 620 University Road, Friday Harbor, WA 98250, USA. E-mail: aquaman@u.washington.edu

†Present address: Department of Zoology, University of Wisconsin, 430 Lincoln Drive, Madison, WI 53706, USA

Resource availability is a second key factor known to influence invasion success and processes that increase or decrease resource availability therefore have strong effects on invasions (Davis *et al.* 2000). Resource pre-emption by native species generates biotic resistance to invasion (Stachowicz *et al.* 1999; Naeem *et al.* 2000; Levine *et al.* 2004). Consequently, physical disturbance can facilitate invasions by reducing competition for limiting resources (Richardson & Bond 1991; Hobbs & Huenneke 1992; Kotanen 1997; Prieur-Richard & Lavorel 2000). In most communities disturbances occur via multiple mechanisms and the disturbances created by different agents vary in their intensity and frequency (D'Antonio *et al.* 1999). Recent empirical (Larson 2003; Hill *et al.* 2005) and theoretical (Higgins & Richardson 1998) studies suggest that not all types of disturbance have equivalent effects on the invasion process. Moreover, most of what we know about the effects of disturbance on invasions comes from short-term experimental studies. It is presently unclear how different disturbance agents influence long-term patterns of invasion.

In order for any invasion to be successful, propagule arrival must coincide with the availability of resources needed by the invading species (Davis *et al.* 2000). Therefore, the interaction between propagule pressure and processes that influence resource availability will ultimately determine invasion success (Brown & Peet 2003; Lockwood *et al.* 2005; Buckley *et al.* 2007). In this study we used the invasion of shallow, subtidal kelp communities in Washington State by the Japanese seaweed *Sargassum muticum* as a study system to better understand the effects of propagule pressure and disturbance on invasion. In a factorial field experiment we manipulated both propagule pressure and disturbance in order to examine how these factors independently and interactively influence *S. muticum* establishment in the short term. We supplement the experimental results with a parameterized integrodifference equation model, which we use to examine how different natural disturbance agents influence the spread of *S. muticum* through the habitat in the longer term. Although a successful invasion clearly requires both establishment and spread of the invader, most studies have looked at just one of these processes (Melbourne *et al.* 2007). We take an integrative approach by employing both a short-term experiment and a longer-term model, allowing us to examine the effects of disturbance and propagule limitation on the entire invasion process.

Methods

STUDY SYSTEM

Our field research was based out of Friday Harbor Laboratories on San Juan Island, Washington State, USA. The field experiment was carried out at a site within the San Juan Islands Marine Preserve network adjacent to Shaw Island, known locally as Point George (48.5549° N, 122.9810° W). Field work was accomplished using SCUBA in shallow subtidal communities.

The native algal community characteristic of sheltered, rocky subtidal habitats in this region is species-rich and structurally complex (see Britton-Simmons 2006 for a more detailed description). In this ecosystem, space is an important limiting resource and in the absence of disturbance there is little or no bare rock available for newly arriving organisms to colonize. This habitat has a diverse fauna of benthic herbivores, including molluscs and sea urchins, that create disturbances by clearing algae from the rocky substrata. The green sea urchin *Strongylocentrotus droebachiensis* is a generalist herbivore that reduces the abundance of native algae and creates relatively large disturbed patches (Vadas 1968; Duggins 1980). In the shallow zone where *S. muticum* is found, the green urchin is highly mobile and often occurs in aggregations (Paine & Vadas 1969; Foreman 1977; Duggins 1983; personal observation). Green urchins avoid areas where *S. muticum* is present because it is not a preferred food resource (Britton-Simmons 2004), but they can be found feeding in uninvaded areas adjacent to existing *S. muticum* populations (personal observation). Green urchins therefore create intermittent but relatively intense disturbances in areas where *S. muticum* is absent and some proportion of these disturbances can potentially be exploited by dispersing *S. muticum* propagules. In contrast, herbivorous benthic molluscs (chitons, limpets and snails) are ubiquitous in the shallow subtidal and unlike sea urchins they are unaffected by the presence of *S. muticum* (Britton-Simmons 2004). Herbivory by individual molluscs creates relatively small-scale disturbances, thereby providing a consistent supply of microsites that can be colonized by newly arriving species, including *Sargassum muticum* (see Appendix S1 in Supplementary Material for more information about mollusc diets).

THE INVADER

Sargassum muticum is a brown alga in the order Fucales that was introduced to Washington State in the early 20th century, probably with shipments of Japanese oysters that were imported for aquaculture beginning in 1902 (Scagel 1956). It is now common in shallow subtidal habitats throughout Puget Sound and the San Juan Islands (Nearshore Habitat Program 2001, personal observation). In the San Juan Islands, *S. muticum* has a pseudoperennial life history. Each holdfast produces as many as 18 laterals in the early spring, each of which can grow as tall as three metres. In late summer to early autumn the laterals senesce and are lost, leaving only the basal holdfast portion of the thallus to overwinter.

Sargassum muticum has a diplontic (uniphasic) life cycle, is monecious, and is capable of selfing. Reproduction typically occurs between late June and late August in our region. During reproduction the eggs of *S. muticum* are released from and subsequently adhere to the outside of small reproductive structures called receptacles. Once fertilized, the resulting embryos remain attached while they develop into tiny germlings (< 200 μm in length) with adhesive rhizoids (Deysher & Norton 1982). Germlings then detach from the receptacle and sink relatively quickly, recruiting in close proximity to the parent plant (Deysher & Norton 1982). Although most recruitment occurs within 5 m of adult plants, recruits have been found as far as 30 m from the nearest adult (Deysher & Norton 1982). Longer distance dispersal probably occurs when plants get detached from the substratum and subsequently become fertile after drifting for some period of time (Deysher & Norton 1982). One distinctive feature of the *S. muticum* invasion is that it is extremely limited in vertical extent. In the San Juan Islands, *S. muticum* is found from the low intertidal to the shallow subtidal zone (Norton 1977; personal observation), from approximately –0.5 m Mean Lower Low Water (MLLW) to –7 m MLLW. However, it is most abundant in the shallow subtidal, from approximately –2 m MLLW to –4 m MLLW. Thus, in areas where *S. muticum* has invaded it forms a narrow band along the shore.

FIELD EXPERIMENT

We used a two-way factorial design manipulating propagule pressure (six levels) and disturbance (two levels) with three replicates per treatment combination. Subtidal plots (30 cm × 30 cm) at a depth of 3–4 m below MLLW were selected so that differences in the identity and abundance of taxa, aspect, and relief were minimized and the plots were randomly assigned to treatments. None of the experimental plots contained *S. muticum* prior to the experiment. However, some *S. muticum* was present at Point George and it was removed prior to the reproductive season in order to prevent contamination of the experimental plots from external sources of *S. muticum* propagules.

The disturbance treatment had two levels: control and disturbed. Control plots were not altered in any way, but they did vary somewhat in how much natural disturbance had occurred in them prior to the experiment (mean = 7.7% of plot area). Plots in the disturbance treatment were scraped down to bare rock so that no visible organisms remained. These two treatments represent extremes in the levels of disturbance that are likely to occur in nature. The unaltered control plots contained a rich assemblage of native species. The disturbed plots were similar in spatial scale to a patch that a small group of urchins might create, but represent an unusually intense disturbance because all native species, including crustose coralline algae (which cover an average of 27.7% of the substratum at this depth), were removed. These treatments maximized our ability to detect an effect of disturbance in our experiment.

Immediately following the imposition of the disturbance treatment (July 2002) the plots were experimentally invaded by suspending 'brooding' *S. muticum* over them. This was accomplished by collecting *S. muticum* from the field and transporting them to the lab where the appropriate ratio of sterile to reproductive tissue (see below) was placed in 30 cm × 30 cm vexar bags. The bags were returned to the field the same day and suspended over the experimental plots for 1 week. Propagule pressure was manipulated by varying the ratio of sterile to reproductive tissue in the bags while holding the total biomass of *S. muticum* tissue constant. The propagule pressure treatment had six levels, corresponding to the following amounts of reproductive tissue (in grams): 0, 50, 100, 175, 250 and 350 (average mass of mature *S. muticum* in this region is 174 g). Based on propagule production–mass relationships derived by Norton & Deysher (1988) for *S. muticum*, we estimate that approximately 5 million propagules were released in each replicate of our highest propagule pressure treatment. We assumed a linear relationship between the mass of adult reproductive tissue and propagule output because we know of no *Sargassum* study that suggests otherwise. Sterile tissue was added to bags as necessary in order to bring the total biomass to 350 g. Reproductive and sterile tissue was mixed in the bags so that the reproductive tissue was well distributed throughout. This experimental manipulation mimics the level of propagule input that would occur in an incipient invasion or if a drifting plant became tangled with attached algae and subsequently released its propagules.

Recruitment of *S. muticum* was quantified by counting the number of *S. muticum* juveniles that were present in the plots 5 months after the experimental invasion, which is the earliest they can reliably be seen in the field. We resurveyed the plots to count the number of *S. muticum* adults present 11 months after the invasion (just prior to reproductive season) and then removed all *S. muticum* from the experimental plots in order to prevent it from spreading.

STATISTICAL ANALYSIS

We analysed the *S. muticum* recruitment data using a two-way ANOVA followed by separate regression analyses on each disturbance treatment. For the control treatment, we performed a multiple regression to determine what proportion of recruitment variation was explained by propagule input and space availability. For the disturbed plots, which did not vary in the amount of available space, we carried out a simple linear regression to determine the impact of propagule input on recruitment. We used the results of these analyses to inform the construction of mechanistic candidate functions for the relationship between propagule input, space availability and recruitment. These candidate functions were compared using differences in the Akaike's information criteria (AIC differences; Burnham & Anderson 2002). We then used model averaging, a form of multimodel inference in which parameter estimates from more than one candidate function are used jointly to describe the data, in order to select a parameterized recruitment function for the *S. muticum* spread model.

The *S. muticum* survivorship data did not conform to the assumptions of ANOVA (even after a number of different transformations) so we used a non-parametric Kruskal–Wallis test to ask whether *S. muticum* survivorship differed in the disturbed and control treatments. We then fitted five different survivorship functions, assuming binomial error, to the data to test whether *S. muticum* survivorship (number of adults per recruit) was density-dependent. Because the Kruskal–Wallis test suggested that survivorship differed significantly between the two disturbance treatments (see Results) we chose to fit the models to those two treatments separately to test for density dependence. In addition to type 1 (linear), type 2 (saturating), and type 3 (sigmoidal) functions, we also fitted a constant survivorship model. These candidate functions were compared using the Akaike's information criterion (AIC differences; Burnham & Anderson 2002).

The numbers of adult *S. muticum* (after 11 months) also violated the assumptions of ANOVA (despite transformations), so we used non-parametric statistics to test two hypotheses: (i) adult density is independent of disturbance treatment (Wilcoxon Signed Ranks Test), and (ii) adult density is independent of propagule pressure treatment (Kruskal–Wallis Test).

MODEL

We used an integrodifference equation (IDE) model to describe the spatial spread of an *S. muticum* population. IDE models assume that the habitat is continuous in space, and that reproduction and dispersal occur in discrete bouts. The depths inhabited by *S. muticum* comprise a relatively narrow vertical band, so the spread of the population was assumed to occur in a one-dimensional habitat. The model follows two state variables through time. $N_t(x)$ is the density of *S. muticum* at a location x along this habitat at time t, and $Z_t(x)$ is the amount of bare rock at x during t. The values for these state variables are determined by functions representing the important ecological processes in this system. *Sargassum muticum* density is determined by the production and recruitment of propagules and by adult survival. Bare rock is created by benthic herbivore disturbances, since herbivores consume native algae and thus alleviate space limitation. The form of our model is then

$$N_{t+1}(x) = sP_t(x)f(P_t(x), Z_t(x)) + rN_t(x), \quad \text{eqn 1}$$

$$Z_{t+1}(x) = (1 - \eta_t(x))gZ_t(x) + \eta_t(x)A. \quad \text{eqn 2}$$

$P_t(x)$ is the number of propagules at location x at the start of year t, and equals the number of propagules produced at x and remaining near their parent plant plus the sum of propagules from all other locations within the habitat (with endpoints a and b) which disperse to x. $P_t(x)$ is governed by the equation $P_t(x) = \int_a^b \omega N_t(y)k(x - y)dy$. Each adult produces ω propagules and their dispersal is described by the function k. The function $f(P_t(x), Z_t(x))$ in equation 1 gives the fraction of propagules which successfully recruit, given that the amount of bare rock at location x equals $Z_t(x)$ and there is an initial input of $P_t(x)$ propagules. Based on data from the experiment, we assume that recruitment function has the form $f(P_t(x),Z_t(x)) = \rho_1(Z_t(x) + \rho_2)^{\rho 5}P_t(x)/[1 + \rho_3(Z_t(x) + \rho_2)^{\rho 5} + \rho_4 P_t(x)^2]$, with values for the ρ_i and methods for fitting this function given in Appendix S2. s and r are fractions of germlings and adults, respectively, that survive to the following year. Parameters for *Sargassum* fecundity and dispersal were attained from the literature (Deysher & Norton 1982; Norton & Deysher 1988) and all other parameter values used in our simulations were estimated from our own field data. The methods and results for fitting parameters are given in Appendix S2.

In equation 2, $\eta_t(x)$ is the proportion of the habitat scraped clear by grazers. If left ungrazed, we assumed that bare rock at a given location experiences geometric decay, with rate g, as it becomes utilized by native algae. The parameter A in equation 2 is a scaling constant representing the size of the habitable area at each point x. We modelled benthic herbivore disturbance in two different ways. First, we constructed a stochastic model for $\eta_t(x)$ based on our understanding of the natural history of the system. Second, we built a more generalized stochastic model for $\eta_t(x)$. In the *S. muticum* system, bare rock is generated in small patches when an area is grazed by molluscs (chitons and limpets), or in larger patches by sea urchin grazing. Both types of disturbance create bare rock for *S. muticum* to potentially exploit, and the disturbance types differ only in their size and spatial distribution. We assumed that the mollusc disturbances are ubiquitous, whereas large urchin-grazed areas are patchily distributed across the habitat. Due to uncertainty in the exact size and frequency of these disturbances, we ran simulations over a very wide range of possible parameter values. In the generalized model for $\eta_t(x)$, we allowed disturbances of any size to occur with any degree of spatial aggregation, rather than requiring large disturbances to be patchy and small ones to be spread throughout the habitat. Our methods for drawing values for $\eta_t(x)$ in these simulations are described in Appendix S3 and summarized in Table C.1 therein.

In our system, native benthic grazers do not eat *S. muticum* adults (Britton-Simmons 2004; personal observation), but it is unknown whether they will consume new *S. muticum* recruits when they are very small (e.g. Sjøtun *et al.* 2007) and hence difficult to avoid ingesting incidentally. Whether or not disturbance events can directly cause mortality of the invader can be very important in determining invasion success (Buckley *et al.* 2007). In our simulations, we therefore considered both the case where *S. muticum* is never eaten by grazers, and the case where *S. muticum* is eaten at the rate $\eta_t(x)$ until it reaches the age of 1 year.

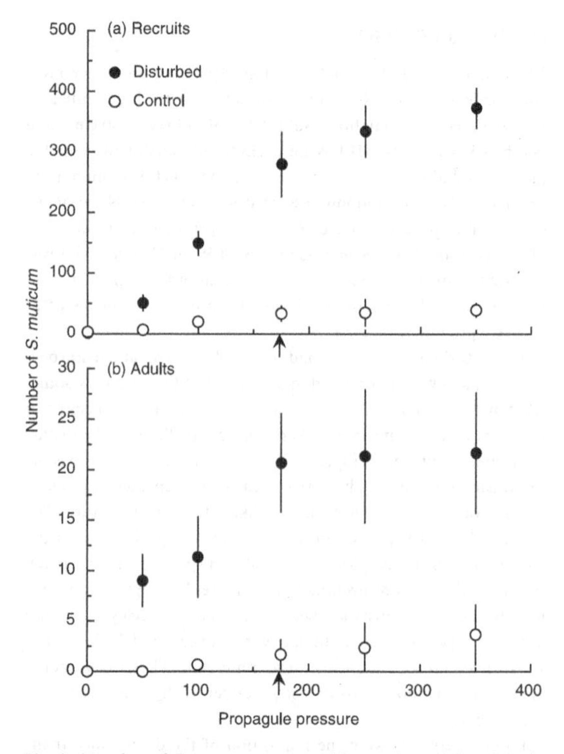

Fig. 1. Number of *Sargassum muticum* (a) recruits and (b) adults in field experiment plots (900 cm²). Propagule pressure is grams of reproductive tissue suspended over experimental plots at beginning of experiment. The average mass of an adult *S. muticum* (174 g) is indicated by an arrow. Data are means ± 1 SE ($n = 3$).

Results

The field experiment showed that recruitment of *S. muticum* was higher in plots that were disturbed compared to control plots (Fig. 1a) suggesting that resource availability limited recruitment. Increasing propagule pressure led to significant increases in average *S. muticum* recruitment in both disturbance treatments (Fig. 1a). Finally, a significant interaction between disturbance and propagule pressure ($F_{5,24} = 3.77$, $P = 0.01$) indicates that the plots in the two disturbance treatments differed in the extent to which they were limited by propagule availability. Multiple regression analysis of the *S. muticum* recruitment data from the control treatment, with space and propagule input as continuous explanatory variables, explained most of the recruitment variability ($R^2 = 0.87$, Fig. 1a). This analysis showed that both space (Fig. 1a, $b = 0.703$, $P < 10^{-4}$) and propagule treatment (Fig. 1a, $b = 0.657$, $P < 10^{-3}$) had strong influences on recruitment in the control treatment. Because there was no variation in space availability in the disturbed treatment, we used simple linear regression analysis to examine the relationship between propagule input and *S. muticum* recruitment in the disturbed treatment (Fig. 1a, $R^2 = 0.84$, $P < 10^{-6}$). The results suggest that in the absence of space limitation propagule input explains most of the variability in *S. muticum* recruitment.

We used these results to create a set of mechanistic candidate functions for the relationship between *S. muticum* recruitment, propagule pressure and space availability (see Appendix S2). The only candidate models supported by the data (AIC differences < 4; Burnham & Anderson 2002) show a type 3 (sigmoidal) relationship between propagule pressure and recruitment, and either a type 2 (saturating) or type 3 relationship between available space and recruitment (Appendix S2, Table B.1). Due to practical constraints on the number of treatments that could be replicated in the field, we have data only on very low available space (control plots) and very high available space (disturbed plots), and insufficient data at intermediate values to resolve the functional relationship between space-limitation and recruitment. We therefore used model averaging (Burnham & Anderson 2002) to combine our parameter estimates for the two supported models and used the resulting function to describe space- and propagule-limitation in recruitment in the simulation model. We also ran simulations using each of the supported recruitment models separately. The results from the two supported models and the averaged model were very similar, so we present results only from the averaged model.

Survivorship (from 5 months to 11 months of age) of *S. muticum* was significantly higher in disturbed plots ($U = 76.5$, $P < 0.05$). Mean survivorship (± 1 SD) in control plots was 3.4% (± 3.8%), compared to 6.1% (± 2.2%) in disturbed plots. Our analysis of survivorship as a function of recruitment density suggests density-independence (Appendix S2, Table B.2), so we used the mean survivorship across all experimental plots as the germling survival rate (s) in our model.

Simulations of the parameterized model under various disturbance regimes reveal several interesting patterns. Using the disturbance scenario with ubiquitous mollusc disturbances and large, patchily distributed urchin disturbances, we found that a single adult *S. muticum* was almost always sufficient to start a successful invasion. This is in agreement with our empirical observation that propagule input always resulted in positive recruitment, even in space-poor control plots. We quantified population growth in our model by reporting the density of *S. muticum* after 100 years, averaged across the invaded area, and we use the length of habitat occupied by *S. muticum* after 100 years as a measure of invasion rate. When we assumed that *S. muticum* was never consumed by benthic herbivores, both the mean *S. muticum* population density and the length of the invaded area increased with both the mean intensity of mollusc grazing and with the size and number of urchin disturbances (Fig. 2, solid lines). Changing the variance in the intensity of mollusc grazing had essentially no effect (not shown). Unless urchin disturbances were extremely large and numerous (top 3 lines, Fig. 2g–j), the mollusc grazing had a much stronger effect on *S. muticum* density than did urchin grazing.

When we assumed that native grazers eat *S. muticum* germlings, *S. muticum* density and the length of habitat invaded still increased with the intensity of mollusc disturbance, as long as molluscs grazed less than 50% of the habitat bare (Fig. 2, dashed lines). Actual mollusc disturbances are typically much smaller than 50% (personal observation). Indeed, we note that if all of the bare rock in the experiment's control plots was attributed to mollusc grazing, the average grazing intensity would be only 7.7%. Within the realistic range of parameter values, then, molluscs facilitate the invasion in the model even when they consume young *S. muticum*.

Urchin disturbances that were few and/or small had little effect on the invasion, but large and numerous urchin disturbances decreased the final *S. muticum* density and the size of the invaded area when grazers consumed new recruits (Fig. 2e–j). *Sargassum muticum* failed to establish when urchin disturbances were both very large (20–50 m of linear habitat scraped bare per disturbance) and extremely abundant (100–200 such disturbances per year). These results are corroborated by the generalized model of disturbance, which showed that when the total proportion of the habitat disturbed per year is held constant smaller disturbances affecting a greater number of locations resulted in the highest final *S. muticum* densities and invaded areas (Appendix S2, Fig. C.1). When these disturbed locations were more clumped in space, this resulted in a slight decrease in the final size of the invaded area.

The treatment effects were still apparent when adults were counted at the end of the experiment (Fig. 1b). Adult *S. muticum* density was higher in the disturbed treatment than in the control treatment ($Z = -3.41$, $P < 0.001$). In addition, adult *S. muticum* density appeared to be positively related to propagule pressure (Fig. 1b, $H_5 = 16.10$, $P = 0.006$), with high propagule pressure resulting in a maximum of between 20 and 25 adults per plot (900 cm^2).

How was the probability of successful invasion influenced by propagule pressure? We defined successful invasion of an experimental plot as the presence of one or more adult *S. muticum* at the end of the experiment (11 months after invasion). We consider this a reasonable way to define invasion success given that reproduction of these adults was imminent (< 1 month away), survivorship is very high at this life-history stage (Appendix S2, Table B.3), and both our model and experimental results indicate that a single individual is capable of establishing a population. We plotted the proportion of plots in each treatment combination that were successfully invaded as a function of propagule pressure (Fig. 3). Because we had only three replicates per treatment combination the probability values were constrained to four possible values (0, 0.33, 0.66, or 1.0). In addition, we tested only six levels of propagule input and therefore have limited capacity to resolve the details of this relationship. Therefore, we did not attempt to fit statistical models to these data. In disturbed plots, invasion was certain even at the lowest level of propagule pressure in our experiment (Fig. 3). However, in control plots the probability of invasion was less than 1 until propagule pressure reached a level of 250 g of reproductive tissue, an amount of tissue greater than the average mass of an adult *S. muticum* (Fig. 3).

Discussion

Our experimental results demonstrate that space- and propagule-limitation both regulate *S. muticum* recruitment. Our finding that *S. muticum* recruitment was positively related to propagule input is similar to those of two previous studies (Parker 2001; Thomsen *et al.* 2006), in which the propagule input of invasive plants was manipulated. In our control

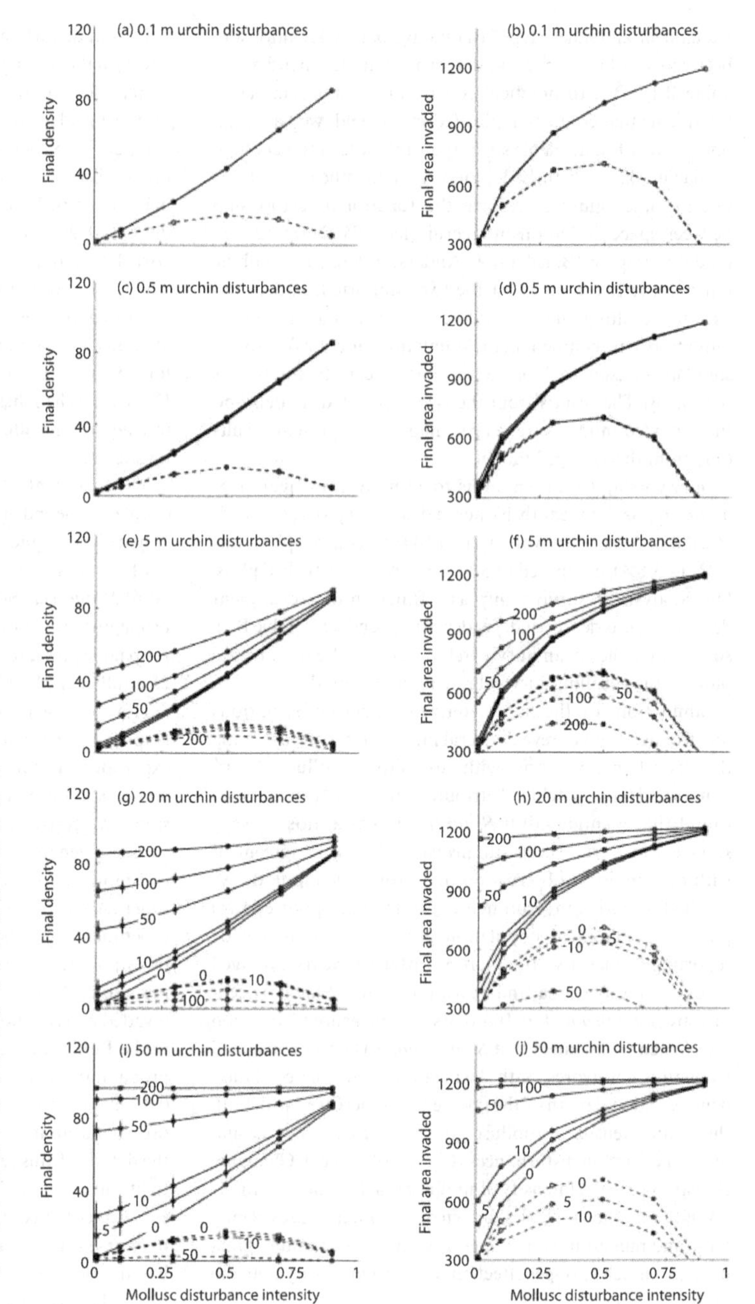

Fig. 2. Simulation results using the mollusc/urchin model for disturbance. The first column (a, c, e, g and i) shows the mean *Sargassum muticum* density (individuals per 900 cm^2) and the second column (b, d, f, h and j) show the length of habitat occupied (metres) after 100 years. Solid lines are the results when native grazers never eat *S. muticum* and dashed lines are results when *S. muticum* recruits (less than 1 year old) are eaten by grazers. The *x*-axis in all plots shows the average proportion of rock scraped bare by molluscs. The number superimposed on each line is the number of urchin disturbances per year (numbers are omitted when the lines overlap completely or are very close together). The mean size of these urchin disturbances increases from the top row (a–b) to the bottom (i–j) and is printed at the top of each graph. Error bars, when large enough to be visible, are ± 1 SE ($n = 100$, as averages were taken across two values for the variance in mollusc intensity with 50 replicates each).

treatment space was limiting, a result that has also been found in previous studies of *S. muticum* recruitment (Deysher & Norton 1982; De Wreede 1983; Sanchez & Fernandez 2006). Consequently, increasing propagule pressure had a relatively weak effect on recruitment in undisturbed plots (Fig. 1a). However, when space limitation was alleviated by disturbing the plots, increasing propagule pressure caused a dramatic increase in recruitment (Fig. 1a). This suggests that in the presence of adequate substratum for settlement, propagule limitation becomes the primary factor controlling *S. muticum* recruitment. These results indicate that *S. muticum* recruitment under natural field conditions will be determined by the interaction between disturbance and propagule input.

Only a few previous studies have investigated the effect of resource supply on the relationship between propagule pressure and recruitment of an introduced species. Although disturbance generally increases invasion success by increasing resource availability (Richardson & Bond 1991; Bergelson

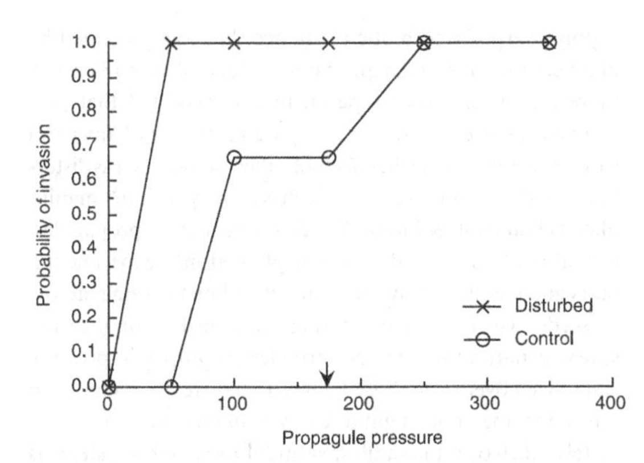

Fig. 3. Probability of invasion as a function of propagule pressure. Probability of invasion is the proportion of plots in each treatment combination ($n = 3$) that contained at least one adult *Sargassum muticum* at the end of the experiment. The average mass of an adult *S. muticum* (174 g) is indicated by an arrow.

et al. 1993; Levin *et al.* 2002; Valentine & Johnson 2003; Clark & Johnston 2005), Parker (2001) found evidence that disturbance reduced Scotch broom (*Cystisus scoparius*) recruitment from seed at all levels of propagule input. This effect occurred because the native flora actually facilitated Scotch broom germination, probably by increasing soil moisture and/or nutrients (Parker 2001). Similarly, Thomsen *et al.* (2006) showed that in the absence of a water addition treatment establishment of an exotic perennial grass was greatly reduced, even at high levels of propagule input. Finally, Valentine & Johnson (2003) found that disturbance facilitated invasion by the introduced kelp *Undaria pinnatifida* even when propagule pressure was high. These studies and our own work provide empirical evidence that the interaction between propagule input and the biotic and abiotic processes that mediate resource availability will be key to understanding patterns of invasion.

The effects of the disturbance and propagule pressure treatments that were manifest in the *S. muticum* recruitment data persisted until the end of the experiment (Fig. 1b). That adult *S. muticum* density was higher in the disturbed treatment than in the control treatment suggests that disturbance may increase the population growth rate of *S. muticum* during the initial stages of the invasion. Natural disturbances that are less intense than our experimental scrapings might have a more modest effect on *S. muticum* density, but our simulation results suggest that even small disturbances can play a major role in facilitating the invasion. Our simulations further suggest that this effect should persist over long time-scales (Fig. 2).

In subtidal habitats both biotic and abiotic disturbances occur, but it is doubtful that they are both relevant to the *S. muticum* invasion in this system. Consumption of algae by the diverse fauna of benthic herbivores in this system (see Methods) is a common and consistent source of disturbance that is likely to be relevant to the *S. muticum* invasion and was therefore the focus of our model. Abiotic disturbances are unlikely to play an important role in this regard because tidal currents are not a substantial cause of algal mortality in this region (Duggins *et al.* 2003) and the inland waters of Puget Sound, the San Juan Islands and the Strait of Georgia are protected from the ocean swells that play a key role on the outer coast of Washington State. Although locally generated storm waves are an important source of disturbance during the winter (Duggins *et al.* 2003), storms during the summer months when *S. muticum* is reproductive are rare.

SIMULATED URCHIN/MOLLUSC DISTURBANCES

In addition to enhancing *S. muticum* recruitment, disturbance increased the survivorship of juvenile *S. muticum*. In our system, the green urchin (*Strongylocentrotus droebachiensis*) creates relatively large disturbed patches and *S. muticum* that recruit to these patches probably benefit from reduced competition with native algae. Unlike other systems where sea urchins feed on both native and non-native algae alike (Valentine & Johnson 2005), green urchins do not consume adult *S. muticum* (Britton-Simmons 2004) although it is possible that they incidentally consume new recruits. Studies in other systems have also reported positive effects of disturbance on the survivorship of non-native species (Gentle & Duggin 1997; Williamson & Harrison 2002). In general, disturbance probably enhances survivorship because it reduces the size or abundance of native species that compete for resources with invaders (Gentle & Duggin 1997; Britton-Simmons 2006). Indeed, our modelling results suggest that even when juvenile survivorship is reduced by herbivory, the net effect of grazers is still usually positive (Fig. 2).

The simulation model suggested that not all disturbance agents have equivalent effects on space-limitation. Small bare patches throughout the habitat facilitated *S. muticum* spread (Fig. 2 and Appendix S3, Fig. C.1) by increasing the amount of bare rock near any given reproductive adult. Molluscs are ubiquitous in these subtidal habitats and although they typically create very small disturbances, the model suggests that this is sufficient for *S. muticum* to successfully invade, even in the absence of other disturbance agents (e.g. urchins and humans).

Urchins create much larger open spaces, but urchin disturbances could not be used by settling propagules unless a reproductive adult happened to be nearby or a long-distance dispersal event occurred. When there are many urchin disturbances in a year, the chance that such a disturbance occurs near an *S. muticum* adult increases and, because long-distance propagule dispersal is rare, this greatly enhances the likelihood that a propagule will reach the disturbed area. Accordingly, small numbers of urchin disturbances in our model did not affect the spread of *S. muticum* (Fig. 2a–d), but numerous and sufficiently large disturbances did (Fig. 2e–j). Washington State is at the southern end of the green urchin's range in the eastern Pacific and at the majority of sites in the San Juan Islands this species is absent or at relatively low abundance. Consequently, molluscs are probably the most important source of disturbance for *S. muticum* in this region; green urchins may be a more important disturbance agent in more northerly portions of its range (where it reaches higher densities). That urchin disturbance was not necessary for successful invasion by *S. muticum* in the model is an important result because *S. muticum* has invaded many areas in this region where urchins are absent. Indeed, urchins avoid areas where *S. muticum* is present (Britton-Simmons 2004) and since this effect was not included in the model, urchin disturbances probably contribute even less to *S. muticum* spread than our simulations suggest.

PROPAGULE PRESSURE AND INVASION SUCCESS

How much invasion risk does a given level of propagule pressure pose? Previous studies have demonstrated a positive relationship between propagule pressure and the establishment success of non-native species (Grevstad 1999; Parker 2001; Ahlroth *et al.* 2003; Cassey *et al.* 2005). However, we know very little about the relationship between establishment probability and propagule pressure or the factors that affect it (Lockwood *et al.* 2005). Possibilities include a linear relationship (Lockwood *et al.* 2005) as well as more complex relationships containing thresholds or other non-linearities (Griffith *et al.* 1989; Ruiz & Carlton 2003; Lockwood *et al.* 2005; Buckley *et al.* 2007). Our experimental results suggest that the relationship is non-linear (Fig. 3). Indeed, all communities in which abiotic factors do not preclude invasion are probably vulnerable to invasion such that above some threshold level of propagule input successful invasion is a virtual certainty. Consequently, this relationship must be nonlinear because by definition it saturates at a probability of one. In our system disturbance appeared to reduce the level of propagule pressure necessary to ensure invasion success. However, even control plots had a high probability of invasion once the level of propagule pressure exceeded that produced by an average adult *S. muticum*. Unfortunately, the limited number of treatment levels in our experiment constrains our ability to resolve the details of this relationship. Nevertheless, in the control treatment there was some evidence of a threshold level of propagule pressure below which invasion was very unlikely to occur (Fig. 3).

Our model reflects what we believe to be the most important factors limiting invasion success (propagule-limitation and competition for space) but other factors we did not include in the model, such as stochastic mortality, density-dependent mortality of adults, competition with native species for resources besides space (e.g. light, Britton-Simmons 2006) and abiotic conditions, could constrain *S. muticum*'s distribution and abundance in the field. Empirical studies have demonstrated the importance of biotic resistance in regulating invasions (see reviews by Levine & D'Antonio 1999; Levine *et al.* 2004) and the community that *S. muticum* is invading is no exception (Britton-Simmons 2006). However, some authors have suggested that propagule pressure has the potential to overcome biotic resistance (D'Antonio *et al.* 2001; Lockwood *et al.* 2005). Levine (2000) found that seed supply overpowered biotic resistance that was generated by plant communities at small spatial scales (18 cm × 18 cm). A more recent terrestrial experiment also reported that propagule pressure was the primary determinant of invasion success, overwhelming the effects of other factors, such as disturbance and resident diversity, which were concurrently manipulated (Von Holle & Simberloff 2005). However, 'propagules' in that study were seedlings transplanted into experimental plots and seedlings may not be regulated by the same factors as seeds, which are the life stage responsible for invasion spread in natural systems. Nevertheless, if propagule pressure can indeed overcome those factors that were not included in our model then one might ask why *S. muticum* has not completely taken over the shallow subtidal zone in this system, as our model predicts under most disturbance regimes. Interestingly, whether *S. muticum* is indeed in the process of doing so is not entirely clear. There are very few areas in the San Juan region where *S. muticum* is completely absent at the appropriate depths (personal observation), yet at many sites *S. muticum* is currently at low abundance and it is unclear whether these sites represent incipient invasions or whether something is inhibiting local population growth.

Conclusions

In our system, neither disturbance nor propagule input alone was sufficient to maximize invasion success (i.e. establishment probability and invader population density). Increasing propagule pressure had relatively little effect on total recruitment in control plots (Fig. 1a), though at high levels it ultimately overcame space limitation and ensured successful invasion (Fig. 3). However, even at high levels of propagule input, final *S. muticum* density was low in the absence of disturbance (Fig. 1b). Based on our experimental results alone, we might have predicted strong effects of both molluscs and urchins on the *S. muticum* invasion in the long term. However, the simulation model suggested that these two natural disturbance agents should have different effects on long-term invasion due to differences in the spatial structure of these disturbances. The model results demonstrate that caution should be exercised when extrapolating the results of short-term disturbance experiments over longer time intervals. In this marine community invasion success was co-regulated by propagule pressure and biotic resistance. Our results underscore the importance of considering interactive effects when making predictions about invasion success.

Acknowledgements

Thanks to Ben Pister, Sam Sublett and Jake Gregg for SCUBA assistance in the field. For helpful discussions that improved this work we thank Timothy Wootton, Cathy Pfister, Greg Dwyer, Joy Bergelson, Mathew Leibold, Spencer Hall and Bret Elderd. Yvonne Buckley, Barney Davies and an anonymous referee provided very helpful comments on an earlier version of the manuscript. The director and staff of Friday Harbor Laboratories provided logistical support and access to laboratory and SCUBA facilities. The field research was funded by a grant to Timothy Wootton from The SeaDoc Society at UC Davis, and both authors were supported by a Graduate Assistance in Areas of National Need Training Grant (P200A040070) during the completion of this work.

References

Ahlroth, P., Alatalo, R., Holopainen, A., Kumpulainen, T. & Suhonen, V. (2003) Founder population size and number of source populations enhance colonization success in waterstriders. *Oecologia*, **137**, 617–620.

Bergelson, J., Newman, J.A. & Floresroux, E.M. (1993) Rates of weed spread in spatially heterogeneous environments. *Ecology*, **74**, 999–1011.

Bossenbroek, J.M., Kraft, C.E. & Nekola, J.C. (2001) Prediction of long-distance dispersal using gravity-models: zebra mussel invasion of inland lakes. *Ecological Applications*, **11**, 1778–1788.

Britton-Simmons, K.H. (2004) Direct and indirect effects of the introduced alga *Sargassum muticum* on benthic, subtidal communities of Washington State, USA. *Marine Ecology Progress Series*, **277**, 61–78.

Britton-Simmons, K.H. (2006) Functional group diversity, resource preemption and the genesis of invasion resistance in a community of marine algae. *Oikos*, **113**, 395 401.

Brown, R.L. & Peet, R.K. (2003) Diversity and invasibility of southern Appalachian plant communities. *Ecology*, **84**, 32 39.

Buckley, Y.M., Bolker, B.M. & Rees, M. (2007) Disturbance, invasion, and reinvasion: managing the weed-shaped hole in disturbed ecosystems. *Ecology Letters*, **10**, 809 817.

Burnham, K.P. & Anderson, D.R. (2002) *Model Selection and Inference: A Practical Information Theoretic Approach*. Springer Publishing, New York, NY.

Cassey, P., Blackburn, T.M., Duncan, R.P. & Lockwood, J.L. (2005) Lessons from the establishment of exotic species: a meta-analytical case study using birds. *Journal of Animal Ecology*, **74**, 250–258.

Clark, G.F. & Johnston, E.L. (2005) Manipulating larval supply in the field: a controlled study of marine invasibility. *Marine Ecology Progress Series*, **298**, 9–19.

D'Antonio, C.M., Dudley, T.L. & Mack, M. (1999) Disturbance and biological invasions: direct effects and feedbacks. *Ecosystems of the World 16: Ecosystems of Disturbed Ground* (ed. L.R. Walker), pp. 413–452. Elsevier.

D'Antonio, C.M., Levine, J. & Thomsen, V. (2001) Ecosytem resistance and the role of propagule supply: a California perspective. *Journal of Mediterranean Ecology*, **2**, 233–245.

Davis, M.A., Grime, J.P. & Thomson, K. (2000) Fluctuating resources in plant communities: a general theory of invasibility. *Journal of Ecology*, **88**, 528–534.

De Wreede, R.E. (1983) *Sargassum muticum* (Fucales, Phaeophyta): regrowth and interaction with *Rhodomela larix* (Ceramiales, Rhodophyta). *Phycologia*, **22** (2), 153–160.

Deysher, L. & Norton, T.A. (1982) Dispersal and colonization in *Sargassum muticum* (Yendo) Fensholt. *Journal of Experimental Marine Biology and Ecology*, **56** (2–3), 179–195.

Drake, J.M., Baggenstos, P. & Lodge, D.M. (2005) Propagule pressure and persistence in experimental populations. *Biology Letters*, **1**, 480–483.

Duggins, D.O. (1980) Kelp beds and sea otters: an experimental approach. *Ecology*, **61**, 447–453.

Duggins, D.O. (1983) Starfish predation and the creation of mosaic patterns in a kelp-dominated community. *Ecology*, **64**, 1610–1619.

Duggins, D.O., Eckman, J.E., Siddon, C.E. & Klinger, T. (2003) Population, morphometric and biomechanical studies of three understory kelps along a hydrodynamic gradient. *Marine Ecology Progress Series*, **265**, 57–76.

Foreman, R.E. (1977) Benthic community modification and recovery following intensive grazing by *Strongylocentrotus droebachiensis*. *Helgoländer wiss. Meeresunters*, **30**, 468–484.

Gentle, C.B. & Duggin, J.A. (1997) *Lantana camara* L. invasions in dry rainforest–open forest ecotones: the role of disturbances associated with fire and cattle grazing. *Australian Journal of Ecology*, **22**, 298–306.

Grevstad, F.S. (1999) Experimental invasions using biological control introductions: the influence of release size on the chance of population establishment. *Biological Invasions*, **1**, 313–323.

Griffith, B., Scott, J.M., Carpenter, J.W. & Reed, C. (1989) Translocation as a species conservation tool: status and strategy. *Science*, **245** (4917), 477–480.

Higgins, S.I. & Richardson, D.M. (1998) Pine invasions in the southern hemisphere: modelling interactions between organism, environment and disturbance. *Plant Ecology*, **135**, 79–93.

Hill, S.J., Tung, P.J. & Leishman, M.R. (2005) Relationships between anthropogenic disturbance, soil properties and plant invasion in endangered Cumberland plain woodland, Australia. *Austral Ecology*, **30**, 775–788.

Hobbs, R.J. & Huenneke, L.F. (1992) Disturbance, diversity, and invasion: implications for conservation. *Conservation Biology*, **6**, 324–337.

Kotanen, P.M. (1997) Effects of experimental soil disturbance on revegetation by natives and exotics on coastal Californian meadows. *Journal of Applied Ecology*, **34**, 631–644.

Larson, D.L. (2003) Native weeds and exotic plants: relationships to disturbance in mixed-grass prairie. *Plant Ecology*, **169**, 317–333.

Levin, P.S., Coyer, J.A., Petrik, R. & Good, T.P. (2002) Community-wide effects of nonindigenous species on temperate rocky reefs. *Ecology*, **83**, 3182–3193.

Levine, J.M. (2000) Species diversity and biological invasions: relating local process to community pattern. *Science*, **288**, 852–854.

Levine, J., Adler, P. & Yelenik, S. (2004) A meta-analysis of biotic resistance to exotic plant invasions. *Ecology Letters*, **7**, 975–989.

Levine, J.M. & D'Antonio, C.M. (1999) Elton revisited: a review of evidence linking diversity and invasibility. *Oikos*, **87**, 15–26.

Lockwood, J.L., Cassey, P. & Blackburn, T. (2005) The role of propagule pressure in explaining species invasions. *Trends in Ecology and Evolution*, **20**, 223–228.

Lonsdale, W.M. (1999) Global patterns of plant invasions and the concept of invasibility. *Ecology*, **80**, 1522–1536.

MacDonald, I., Loope, L., Usher, M. & Hamann, O. (1989) Wildlife conservation and the invasion of nature reserves by introduced species: a global perspective. *Biological Invasions: A global perspective* (eds J.A. Drake, H.A. Mooney, F. di Castri, R.H. Groves, F.J. Kruger, M. Rejmanek & M. Williamson), pp. 215 255. John Wiley & Sons, Chichester, UK.

Mack, M., Simberloff, D., Lonsdale, W., Evans, H., Clout, M. & Bazzaz, F. (2000) Biotic invasions: causes, epidemiology, global consequences, and control. *Ecological Applications*, **10**, 689 710.

Melbourne, B.A., Cornell, H.V., Davies, K.F., Dugaw, C.J., Elmendorf, S., Freestone, A.L., Hall, R.J., Harrison, S., Hastings, A., Holland, M., Holyoak, M., Lambrinos, J., Moore, K. & Yokomizo, H. (2007) Invasion in a heterogeneous world: resistance, coexistence or hostile takeover? *Ecology Letters*, **10**, 77–94.

Naeem, S., Knops, J., Tilman, D., Howe, K., Kennedy, T. & Gale, S. (2000) Plant diversity increases resistance to invasion in the absence of covarying extrinsic factors. *Oikos*, **91**, 97–108.

Nearshore Habitat Program (2001) *The Washington State ShoreZone Inventory*. Washington State Department of Natural Resources, Olympia, WA.

Norton, T.A. (1977) Ecological experiments with *Sargassum muticum*. *Journal of the Marine Biology Association of the United Kingdom*, **57**, 33–43.

Norton, T.A. & Deysher, L.E. (1988) The reproductive ecology of *Sargassum muticum* at different latitudes. *Reproduction, Genetics and Distributions of Marine Organisms: 23rd European Marine Biology Symposium, School of Biological Sciences, University of Wales, Swansea* (eds J.S. Ryland & P.A. Tyler), pp. 147–152. Olsen & Olsen, Fredensborg, Denmark.

Paine, R.T. & Vadas, R.L. (1969) The effects of grazing by sea urchins, *Strongylocentrotus* spp., on benthic algal populations. *Limnology and Oceanography*, **14**, 710–719.

Panetta, F.D. & Randall, R.P. (1994) An assessment of the colonizing ability of *Emex australis*. *Australian Journal of Ecology*, **19**, 76–82.

Parker, I.M. (2001) Safe site and seed limitation in *Cystisus scoparius* (Scotch Broom): Invasibility, disturbance, and the role of cryptograms in a glacial outwash prairie. *Biological Invasions*, **3**, 323–332.

Pimentel, D., Zuniga, R. & Morrison, D. (2005) Update on the environmental and economic costs associated with alien-invasive species in the United States. *Ecological Economics*, **52**, 273–288.

Prieur-Richard, A. & Lavorel, S. (2000) Invasions: the perspective of diverse plant communities. *Austral Ecology*, **25**, 1–7.

Richardson, D.M. & Bond, W.J. (1991) Determinants of plant distribution: evidence from pine invasions. *American Naturalist*, **137**, 639–668.

Ruiz, G.M. & Carlton, J.T. (2003) Invasion vectors: a conceptual framework for management. *Invasive Species: Vectors and Management Strategies* (eds G.M. Ruiz & J.T. Carlton), pp. 459–504. Island Press.

Sanchez, I. & Fernandez, C. (2006) Resource availability and invasibility in an intertidal macroalgal assemblage. *Marine Ecology Progress Series*, **313**, 85–94.

Scagel, R.F. (1956) Introduction of a Japanese alga, *Sargassum muticum*, into the Northeast Pacific. *Washington Department of Fisheries, Fisheries Research Papers*, **1**, 1–10.

Simberloff, D. (1989) Which insect introductions succeed and which fail? *Biological Invasions: A global perspective, SCOPE 37* (eds J.A. Drake, H.A. Mooney, F. di Castri, R.H. Groves, F.J. Kruger, M. Rejmanek and M. Williamson), pp. 61–75. John Wiley & Sons, Chichester, UK.

Sjøtun, K., Eggereide, S.F. & Høisaeter, T. (2007) Grazer-controlled recruitment of the introduced *Sargassum muticum* (Phaeophyceae, Fucales) in northern Europe. *Marine Ecology Progress Series*, **342**, 127–138.

Stachowicz, J., Whitlatch, R. & Osman, R. (1999) Species diversity and invasion resistance in a marine ecosystem. *Science*, **286**, 1577–1579.

Thomsen, M.A., D'Antonio, C.M., Suttle, K.B. & Sousa, W.P. (2006) Ecological resistance, seed density and their interactions determine patterns of invasion in a California grassland. *Ecology Letters*, **9**, 160–170.

Travis, J.M.J., Hammershoj, M. & Stephenson, C. (2005) Adaptation and propagule pressure determine invasion dynamics: insights from a spatially explicit model for sexually reproducing species. *Evolutionary Ecology Research*, **7**, 37–51.
Vadas, R.L. (1968) *The ecology of* Agarum *and the kelp bed community.* PhD Dissertation, University of Washington.
Valentine, J.P. & Johnson, C.R. (2003) Establishment of the introduced kelp *Undaria pinnatifida*. Tasmania depends on disturbance to native algal assemblages. *Journal of Experimental Marine Biology and Ecology*, **265**, 63–90.
Valentine, J.P. & Johnson, C.R. (2005) Persistence of the exotic kelp *Undaria pinnatifida* does not depend on sea urchin grazing. *Marine Ecology Progress Series*, **285**, 43–55.
Veltman, C.J., Nee, S. & Crawley, M.J. (1996) Correlates of introduction success in exotic New Zealand birds. *American Naturalist*, **147**, 542–557.
Von Holle, B. & Simberloff, D. (2005) Ecological resistance to biological invasion overwhelmed by propagule pressure. *Ecology*, **86**, 3212–3218.
Williamson, M. (1989) Mathematical models of invasio. *Biological Invasions: A global perspective* (ed. by J.A. Drake, H.A. Mooney, F. di Castri, R.H. Groves, F.J. Kruger, M. Rejmanek and M. Williamson), pp. 329–350. John Wiley & Sons, Chichester, UK.
Williamson, M. (1996) *Biological Invasions*. Chapman & Hall, London, UK.
Williamson, J. & Harrison, S. (2002) Biotic and abiotic limits to the spread of exotic revegetation species. *Ecological Applications*, **12**, 40–51.

Received 12 June 2007; accepted 1 October 2007
Handling Editor: Jonathan Newman

Supplementary material

The following supplementary material is available for this article:

Appendix S1. Detailed diet information for benthic, subtidal mollusc species.

Appendix S2. Model parameter values and functions.

Appendix S3. Models for disturbance.

This material is available as part of the online article from: http://www.blackwell-synergy.com/doi/abs/10.1111/j.1365-2745.2007.01319.x
(This link will take you to the article abstract).

Please note: Blackwell Publishing is not responsible for the content or functionality of any supplementary materials supplied by the authors. Any queries (other than missing material) should be directed to the corresponding author for the article.

CHAPTER 21

PEA3：Ganci et al. (2012)

Remote Sensing of Environment 119 (2012) 197–207

Contents lists available at SciVerse ScienceDirect

Remote Sensing of Environment

journal homepage: www.elsevier.com/locate/rse

An emergent strategy for volcano hazard assessment: From thermal satellite monitoring to lava flow modeling

Gaetana Ganci [a,*], Annamaria Vicari [a], Annalisa Cappello [a,b], Ciro Del Negro [a]

[a] *Istituto Nazionale di Geofisica e Vulcanologia, Sezione di Catania, Osservatorio Etneo, Italy*
[b] *Dipartimento di Matematica e Informatica, Università di Catania, Italy*

ARTICLE INFO

Article history:
Received 8 September 2011
Received in revised form 21 December 2011
Accepted 23 December 2011
Available online 26 January 2012

Keywords:
Etna volcano
Infrared remote sensing
Numerical simulation
GIS
Lava hazard assessment

ABSTRACT

Spaceborne remote sensing techniques and numerical simulations have been combined in a web-GIS framework (LAV@HAZARD) to evaluate lava flow hazard in real time. By using the HOTSAT satellite thermal monitoring system to estimate time-varying TADR (time averaged discharge rate) and the MAGFLOW physics-based model to simulate lava flow paths, the LAV@HAZARD platform allows timely definition of parameters and maps essential for hazard assessment, including the propagation time of lava flows and the maximum run-out distance. We used LAV@HAZARD during the 2008–2009 lava flow-forming eruption at Mt Etna (Sicily, Italy). We measured the temporal variation in thermal emission (up to four times per hour) during the entire duration of the eruption using SEVIRI and MODIS data. The time-series of radiative power allowed us to identify six diverse thermal phases each related to different dynamic volcanic processes and associated with different TADRs and lava flow emplacement conditions. Satellite-derived estimates of lava discharge rates were computed and integrated for the whole period of the eruption (almost 14 months), showing that a lava volume of between 32 and 61 million cubic meters was erupted of which about 2/3 was emplaced during the first 4 months. These time-varying discharge rates were then used to drive MAGFLOW simulations to chart the spread of lava as a function of time. TADRs were sufficiently low (<30 m^3/s) that no lava flows were capable of flowing any great distance so that they did not pose a hazard to vulnerable (agricultural and urban) areas on the flanks of Etna.

1. Introduction

Etna volcano, Italy, is characterized by persistent activity, consisting of degassing and explosive phenomena at its summit craters, as well as frequent flank eruptions. All eruption typologies can give rise to lava flows, which are the greatest hazard presented by the volcano to inhabited and cultivated areas (Behncke et al., 2005). The frequent eruptions of Mt Etna and hazard they pose represent an excellent opportunity to apply both spaceborne remote sensing techniques for thermal volcano monitoring, and numerical simulations for predicting the area most likely to be inundated by lava during a volcanic eruption. Indeed, Etna has witnessed many tests involving the application of space-based remote sensing data to detect, measure and track the thermal expression of volcanic effusive phenomena (e.g. Ganci et al., 2011b; Harris et al., 1998; Wright et al., 2004), as well as a number of lava flow emplacement models (e.g. Crisci et al., 1986; Harris & Rowland, 2001; Vicari et al., 2007). These efforts have allowed key at-risk areas to be rapidly and appropriately identified (Ganci et al., 2011b; Wright et al., 2008).

Over the last 25 years, satellite measurements in the thermal infrared have proved well suited to detection of volcanic thermal phenomena (e.g. Francis & Rothery, 1987; Harris et al., 1995) and to map the total thermal flux from active lava flows (e.g. Flynn et al., 1994; Harris et al., 1998). High spatial resolution data collected by Landsat and ASTER have been employed for the thermal analysis of active lava flows (Hirn et al., 2007; Oppenheimer, 1991), lava domes (Carter & Ramsey, 2010; Kaneko et al., 2002; Oppenheimer et al., 1993), lava lakes (Harris et al., 1999; Wright et al., 1999), and fumarole fields (Harris & Stevenson, 1997; Pieri & Abrams, 2005). Lower spatial, but higher temporal, resolution sensors, such the Advanced Very High Resolution Radiometer (AVHRR) and the MODerate Resolution Imaging Spectro-radiometer (MODIS), have also been widely used for infrared remote sensing of volcanic thermal features, as has GOES (e.g. Rose & Mayberry, 2000). The high temporal resolution (15 min) offered by the Spinning Enhanced Visible and Infrared Imager (SEVIRI), already employed for the thermal monitoring of effusive volcanoes in Europe and Africa (Hirn et al., 2008), has recently been exploited to estimate lava discharge rates for eruptive events of short duration (a few hours) at Mt Etna (Bonaccorso et al., 2011a, 2011b). These data have also proved capable of detecting, measuring and tracking volcanic thermal phenomena, despite the fact that the volcanic thermal phenomena are usually much smaller than the

* Corresponding author at: Istituto Nazionale di Geofisica e Vulcanologia (INGV), Piazza Roma 2, 95123 Catania, Italy. Tel.: +39 095 7165800; fax +39 095 435801.
E-mail address: gaetana.ganci@ct.ingv.it (G. Ganci).

0034-4257/$ – see front matter © 2012 Elsevier Inc. All rights reserved.
doi:10.1016/j.rse.2011.12.021

nominal pixel sizes of 1 to 4 km (Ganci et al., 2011a, 2011b; Vicari et al., 2011b). Great advances have also been made in understanding the physical processes that control lava flow emplacement, resulting in the development of a range of tools allowing the assessment of lava flow hazard. Many methods have been developed to simulate and predict lava flow paths and run-out distance based on various simplifications of the governing physical equations and on analytical and empirical modeling (e.g.: Hulme, 1974; Crisci et al., 1986; Young & Wadge, 1990; Miyamoto & Sasaki, 1997; Harris & Rowland, 2001; Costa & Macedonio, 2005; Del Negro et al., 2005). In its most simple form, such forecasting may involve the application of volcano-specific empirical length/effusion rate relationships to estimate flow length (e.g. Calvari & Pinkerton, 1998). At their most complex, such predictions may involve iteratively solving a system of equations that characterize the effects that cooling-induced changes in rheology have on the ability of lava to flow downhill, while taking into account spatial variations in slope determined from a digital elevation model (e.g. Del Negro et al., 2008).

On Etna, the MAGFLOW Cellular Automata model has successfully been used to reproduce lava flow paths during the 2001, 2004 and 2006 effusive eruptions (Del Negro et al., 2008; Herault et al., 2009; Vicari et al., 2007). More recently, it has been applied to consider the impact of hypothetical protective barrier placement on lava flow diversion (Scifoni et al., 2010) and has been used for the production of a hazard map for lava flow invasion on Etna (Cappello et al., 2011a, 2011b). MAGFLOW is based on a physical model for the thermal and rheological evolution of the flowing lava. To determine how far lava will flow, MAGFLOW requires constraint of many parameters. However, after chemical composition, the instantaneous lava output at the vent is the principal parameter controlling the final dimensions of a lava flow (e.g. Harris & Rowland, 2009; Pinkerton & Wilson, 1994; Walker, 1973). As such, any simulation technique that aims to provide reliable lava flow hazard assessments should incorporate temporal changes in discharge rate into its predictions in a timely manner.

Satellite remote sensing provides a means to estimate this important parameter in real-time during an eruption (e.g. Harris et al., 1997; Wright et al., 2001). A few attempts to use satellite-derived discharge rates to drive numerical simulations have already been made (e.g. Herault et al., 2009; Vicari et al., 2009; Wright et al., 2008). These modeling efforts all used infrared satellite data acquired by AVHRR and/or MODIS to estimate the time-averaged discharge rate (TADR) following the methodology of Harris et al. (1997). Recently, we developed the HOTSAT multiplatform system for satellite infrared data analysis, which is capable of managing multispectral data from different sensors as MODIS and SEVIRI to detect volcanic hot spots and output their associated radiative power (Ganci et al., 2011b). Satellite-derived output for lava flow modeling purposes has already been tested for the 12–13 January 2011 paroxysmal episode at Mt Etna by Vicari et al., 2011b. Here we present a new web-based GIS framework, named LAV@HAZARD, which integrates the HOTSAT system for satellite-derived discharge rate estimates with the MAGFLOW model to simulate lava flow paths. As a result, LAV@HAZARD now represents the central part of an operational monitoring system that allows us to map the probable evolution of lava flow-fields while the eruption is ongoing. Here we describe and demonstrate the operation of this LAV@HAZARD using a retrospective analysis of Etna's 2008–2009 flank eruption. This eruption was exceptionally well documented by a variety of monitoring techniques maintained by INGV-CT including spaceborne thermal infrared measurements. Within this operational role, HOTSAT is first used to identify thermal anomalies due to active lava flows and to compute the TADR. This is then used to drive MAGFLOW simulations, allowing us to effectively simulate the advance rate and maximum length for the active lava flows. Because satellite-derived TADRs can be obtained in real times and simulations spanning several days of eruption can be calculated in a few minutes, such a combined approach has the potential to provide timely predictions of the areas likely to be inundated with lava, which can be updated in response to changes of the eruption conditions as detected by the current image conditions. If SEVIRI data are used, simulations can be updated every 15 min. Our results thus demonstrate how LAV@HAZARD can be exploited to produce realistic lava flow hazard scenarios and for helping local authorities in making decisions during a volcanic eruption.

2. HOTSAT satellite monitoring system

HOTSAT is an automated system that ingests infrared data acquired by MODIS and SEVIRI data. The decision to employ both SEVIRI (Govaerts et al., 2001) and MODIS (Running et al., 1994) sensors is due to the advantages furnished by their different resolutions. In particular, the high temporal resolution offered by SEVIRI enables almost continuous monitoring (i.e. up to four times per hour) of volcanic thermal features allowing us to measure short and rapidly evolving eruptive phenomena. On the other hand, the higher spatial resolution (1 km, as opposed to 3 km, pixels), the good spectral resolution and the high signal to noise ratio of MODIS permit to detect less intense thermal anomalies and to locate them with more detail. Moreover, the "fire" channel of MODIS (channel 21, 3.9 μm) is designed with a higher saturation temperature of about 500 K (with respect to SEVIRI channel 4, which is also located at 3.9 μm and saturated at about 335 K).

2.1. Hot spot detection

The automatic detection in satellite images of hot spots (thermal anomalies that possibly relate to dynamic volcanic processes) is a nontrivial question, since an appropriate threshold radiance value must be chosen to reveal the pixels containing thermal anomalies. Traditional algorithms rely on simple threshold tests, and open questions arise from the right choice of the thresholds based on latitude, season and time of the day. To overcome limits due to a fixed threshold, we applied a contextual algorithm to both MODIS and SEVIRI images. As starting point, we take the contextual approach of Harris et al. (1995), where a non-volcanic area is defined in order to calculate a threshold from within the image. The algorithm computes for each pixel the difference (ΔT_{diff}) between brightness temperature in mid-infrared (MIR: 3.9 μm) and thermal infrared (TIR: 10–12 μm), and the spatial standard deviation (SSD) of ΔT_{diff}. A threshold (SSD_{max}) is then defined as the maximum of $SSD(\Delta T_{diff})$ from the non-volcanic area within the image.

To locate thermal anomalies automatically, we introduced a two-step approach (Ganci et al., 2011b). In a first step, all the pixels belonging to the "volcanic" area of the image are scanned and those having a value of $SSD(\Delta T_{diff})$ greater than SSD_{max} are classified as "potentially" hot. In a second step, all the "potentially" hot pixels and their neighbors are analyzed to assess whether the potential hot spot is valid. The detection is considered valid if at least one of two conditions was met:

$$T_{3.9\mu m} - \mathtt{min}\left(T_{3.9\mu m}\right) > \mathtt{MaxVar}\left(T_{3.9\mu m}\right) \quad (1)$$

$$T_{3.9\mu m} > \mathtt{mean}\left(T_{3.9\mu m}\right) + n^{*}\mathtt{std}\left(T_{3.9\mu m}\right) \quad (2)$$

where $T_{3.9\mu m}$ is the 3.9 μm brightness temperature, and $MaxVar(T_{3.9\mu m})$ and $std(T_{3.9\mu m})$ are respectively the maximum variation and the standard deviation of the 3.9 μm brightness temperature computed in the non-volcanic area of the image. Parameter n controls how much the MIR pixel-integrated temperature deviates from the mean value. Following this procedure, all the computations are based on dynamic thresholds calculated for the image in hand.

2.2. Heat flux and TADR conversion

HOTSAT next calculates the radiative power for all "hot" pixels. To do this, we applied the MIR radiance conversion of Wooster et al. (2003), an approach that allows an estimate of the radiative power from a sub-pixel hot spot using an approximation of Planck's Law. For each hot spot pixel, the radiative power from all hot thermal components is calculated by combining Stefan–Boltzmann law and Planck's Law, obtaining:

$$FRP_{MIR} = \frac{A_{sampl}\varepsilon\sigma}{a\varepsilon_{MIR}} L_{MIR,h} \quad (3)$$

where, Q_{pixel} is the radiative power radiated by the thermally anomalous pixel [W], A_{sampl} is the ground sampling (i.e. pixel) area [m^2], ε is the emissivity, σ is the Stefan–Boltzmann constant [$5.67\times10^{-8}\,J\,s^{-1}\,m^{-2}\,K^{-4}$], $L_{MIR,h}$ and ε_{MIR} are the hot pixel spectral radiance and surface spectral emissivity in the appropriate MIR spectral band, and constant a [$W\,m^{-4}\,sr^{-1}\,\mu m^{-1}\,K^{-4}$] is determined from empirical best-fit relationships. We use the value of a as defined by Wooster et al. (2005) for measurements at 3.9 μm. From Eq. (3), the radiative power is proportional to the calibrated radiance associated with the hot part of the pixel computed as the difference between the observed hot spot pixel radiance in the MIR channel and the background radiance, which would have been observed at the same location in the absence of thermal anomalies.

Harris et al. (1997) showed that the total radiative power measured from satellite infrared data can be converted to time-averaged discharge rate (TADR) using:

$$TADR = \frac{Q}{\rho(C_P\Delta T + C_L\Delta\varphi)} \quad (4)$$

where Q is the total thermal flux obtained summing up the radiative power computed for each hot spot pixel, ρ is the lava density, C_P is the specific heat capacity, ΔT is the eruption temperature minus temperature at which flow stops, C_L is the latent heat of crystallization, and ΔΦ is the volume percent of crystals that form while cooling through ΔT. It is worth noting that the conversion from heat flux to volume flux depends on many lava parameters (such as density, specific heat capacity, eruption temperature, etc.) and has to be determined depending on flow conditions (insulation, rheology, slope, crystallinity, etc.) (Harris et al., 2010). As a result, the conversion has generally been reduced to a linear best-fit relation with the form TADR = xQ, where x needs to be set appropriately (Harris & Baloga, 2009; Wright et al., 2001). Given that we cannot fix a single value to characterize this conversion, the most reasonable solution is to use a range of possible values. We thus defined a range of solutions by adopting the extreme values (Table 1) found by Harris et al. (2000, 2007), to be appropriate for calibrating this technique for Etna lavas.

Table 1
Lava parameter values used to convert satellite thermal data to TADR at Mt Etna (following Harris et al., 2000, 2007) and to run MAGFLOW simulations.

Parameter	Description	Value
ρ	Dense rock density	2600 kg m^{-3}
C_p	Specific heat capacity	1150 J kg^{-1} K^{-1}
ϖ	Vesicularity	10–34%
ε	Emissivity	0.9
T_s	Solidification temperature	1173 K
T_e	Extrusion temperature	1360 K
ΔT	Difference between eruption temperature and temperature at which flow is no longer possible	100–200 K
ΔΦ	Crystallization in cooling through ΔT	30–54%
C_L	Latent heat of crystallization	3.5×10^5 J kg^{-1}

3. MAGFLOW lava flow simulator

MAGFLOW is a physics-based numerical model for lava flow simulations based on a Cellular Automaton (CA) approach developed at INGV-Catania (Del Negro et al., 2008; Vicari et al., 2007). The MAGFLOW cellular automaton has a two-dimensional structure with cells described by five scalar quantities: ground elevation, lava thickness, heat quantity, temperature, and amount of solidified lava. The system evolution is purely local, with each cell evolving according to its present status and the status of its eight immediate neighbors (i.e. its Moore neighborhood). In this way, the CA can produce extremely complex structures using simple and local rules. The domain (automaton size) needs to be large enough to include the expected maximum extent of the lava flow, and is decomposed into square cells whose width matches the resolution of the Digital Elevation Model (DEM) available for the area. Lava thickness varies according to lava influx between source cells and any neighboring cells. Lava flux between cells is determined according to the height difference in the lava using a steady-state solution for the one-dimensional Navier–Stokes equations for a fluid with Bingham rheology. Conservation of mass is guaranteed both locally and globally (Dragoni et al., 1986).

MAGFLOW uses a thermo-rheological model to estimate the point at which the temperature of a cell drops below a given solidus temperature T_s. This defines the temperature at which lava stops flowing. A corresponding portion of the erupted lava volume remaining in the cell is converted to solid lava. This specifies the total height of the cell but not to the amount of fluid that can move. For our purposes, the thickness of lava remaining in each cell is initially set to zero. Lava flow is then discharged at a certain rate from a cell (or group of cells) corresponding to the vent location in the DEM. The thickness of lava at the vent cell increases by a rate calculated from the volume of lava extruded during each time interval (where the flow rate from vent can change in time). When the thickness at the vent cell reaches a critical level, the lava spreads into neighboring cells. Next, whenever the thickness in any cell exceeds the critical thickness, lava flows into the adjacent cells.

To produce a dynamic picture of probable lava flow paths, MAGFLOW requires constraint of many parameters. These include: (i) a knowledge of the chemical composition of the lava (this places constraints on the eruption temperature of the lava, the relationships between lava temperature and viscosity, and temperature and yield strength, all of which are used to compute the rate at which the lava solidifies), (ii) a digital representation of the topography over which the lava is to be emplaced, (iii) the location of the eruptive vents, and (iv) an estimate of the lava discharge rate. However, it has been well established, and confirmed many times in the literature (see Harris & Rowland, 2009 for review) that, for a given chemical composition, a higher discharge rate allows a volume of lava to spread over a greater area before it solidifies than if fed at a lower discharge rate. As such, simulations that take into account the way in which discharge rate changes during an eruption, and how this influences lava spreading as a function of time, are of special interest, particularly as effusion rates can vary even over relatively short time scales (Harris et al., 2007).

We have just submitted a paper (Bilotta et al., unpublished results) in which the results of a sensitivity analysis of the MAGFLOW Cellular Automaton model are presented. In this companion paper we carry out a sensitivity analysis of the physical and rheological parameters that control the evolution function of the automaton. The results confirm that, for a given composition, discharge rates strongly influence the modeled emplacement. Indeed, to obtain more accurate simulations (all giving the same lava volume), it is better to input into the model a continuous time-varying discharge rate, even if with moderate errors, rather than sparse but accurate measurements. Such records can be created using satellite remote sensing, and can be made available within hours, perhaps minutes, of satellite overpass, especially when using data acquired by low spatial, but high temporal resolution sensors, such as MODIS and SEVIRI.

4. LAV@HAZARD web-GIS framework

We integrated the HOTSAT system and the MAGFLOW model using a web-based Geographical Information System (GIS) framework, named LAV@HAZARD (Fig. 1). In a companion paper, Vicari et al., (2011a) review the technical aspects of the framework. Here we detail the system functionality for tracking an on-going effusive eruption using Etna most recent flank event as an illustrative case-study. By using the satellite thermal monitoring system to estimate time-varying discharge rates and the physics-based model to simulate lava flow paths, our combined approach allows timely definition of parameters essential for hazard monitoring purposes, such as the time of propagation of lava flow fronts, maximum run-out distance, and area of inundation. The choice of the web architecture allows remote control of the whole platform in a rapid and easy way and complete functionality for real-time lava flow hazard assessment. To do this, we exploit the web-based mapping service provided by Google Maps. Due to its wide-spread use, and due to the fact that it allows a high degree of customization through a number of utilities called Application Programming Interfaces (APIs), Google provides a perfect foundation for our system. The wide array of APIs furnished by Google Maps allows in a very simple way to embed and manipulate maps, and to place over them different information layers.

LAV@HAZARD consists of four modules regarding monitoring and assessment of lava flow hazard at Etna: (i) satellite-derived output by HOTSAT (including time-space evolution of hot spots, radiative power, discharge rate, etc.), (ii) lava flow hazard map visualization, (iii) a database of MAGFLOW simulations of historic lava flows, and (iv) real-time scenario forecasting by MAGFLOW. As part of the satellite module, MODIS and SEVIRI images are automatically analyzed by HOTSAT, which promptly locates the thermally anomalous pixels. Next, the heat flux from the anomalous pixels is calculated, which is converted to lava discharge rates. Using SEVIRI, these can be calculated, up-date on (and added to) the LAV@HAZARD database up to four times per hour. The satellite-derived discharge rate estimates are then used in the scenario forecasting module as input parameters to MAGFLOW. Because HOTSAT provides minimum and maximum estimates for TADR, two corresponding (end member) lava flow simulations are produced by MAGFLOW. Every time the satellite obtains a new image of Etna volcano, the TADR is updated, and a new pair of simulations is produced.

Although our original implementation of MAGFLOW was intended for serial execution on standard CPUs (Computer Processing Units), the cellular automaton paradigm displays a very high degree of parallelism that makes it suitable for implementation on parallel computing hardware. Thus, in LAV@HAZARD, MAGFLOW has been implemented on Graphic Processing Units (GPUs), these offering very high performances in parallel computing with a total cost of ownership that is significantly inferior to that of traditional computing clusters of equal performance. The porting of MAGFLOW from the original serial code to the parallel computational platforms was accomplished by Bilotta et al. (2011) using CUDA (Compute Unified Device Architecture), a parallel computing architecture provided by NVIDIA Corporation for the deployment of last generations of GPUs as high-performance computing hardware. The benefit of running on GPUs, rather than on CPUs, depends on the extent and duration of the simulated event; while for large, long-running simulations, the GPU can be 70-to-80 times faster, for short-lived eruptions (emplacing lava units of limited areal extent) the increase in speed obtained is between 40-and-50 times. This means that running MAGFLOW on GPUs provides a simulation spanning several days of eruption in a few minutes. In this way, predictions of likely lava flow paths can be promptly produced in a timely fashion and faster than the rate of TADR update (15 min).

Automatic updating of lava flow hazard scenarios and the remote control of HOTSAT and MAGFLOW on the web, make LAV@HAZARD a helpful tool to validate and adjust/refine our output in real time. The

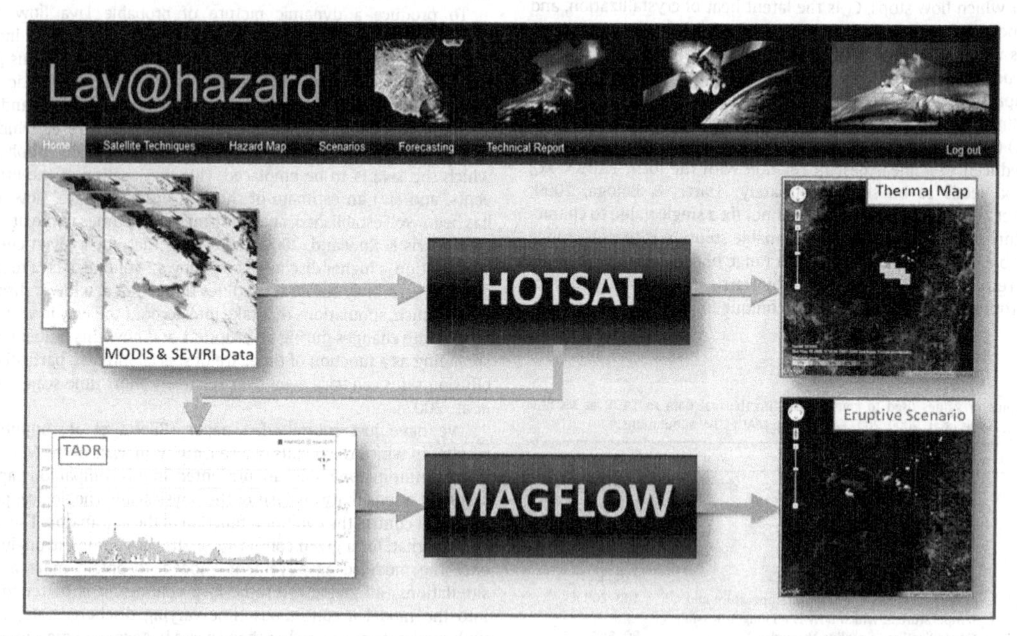

Fig. 1. Flow diagram of the Web-GIS framework LAV@HAZARD. The sketch shows the main elements and products of HOTSAT and MAGFLOW. Thermal maps and TADRs are obtained from MODIS and SEVIRI data. The satellite-derived eruption rates are employed to produce eruptive scenarios.

results of running our web-GIS system during Etna's 2008–2009 eruption are described next.

5. Case study: 2008–2009 Etna eruption

On 10 May 2008 Etna entered an explosive phase, followed by the next effusive eruption 3 days later. Beginning at ~14:00 GMT on 10 May a new vent opened at the eastern base of the South East Crater (SEC) (see Fig. 2). This vent fed intense lava fountaining (where stronger gas jets propelled lava fragments to heights of several hundred meters) characterized by an unusually high rate that in only 4–5 h formed a large lava field. The effusive flank eruption began on 13 May 2008 with the opening of a system of eruptive fissures that propagated SE from the summit craters toward the western wall of the Valle del Bove (VdB). Lava fountains erupted from a fissure extending between 3050 and 2950 m.a.s.l. In 2 h, the eruptive fissure propagated downslope and southeastward, reaching a minimum elevation of 2620 m.a.s.l. (Bonaccorso et al., 2011a). Meanwhile, Strombolian activity migrated from the upper segment of the fissure to its lowest portion. Here, a lava flow erupted at high rates from two main vents (V_1 and V_2 in Fig. 2) and rapidly expanded in the VdB to reach a maximum distance of 6.4 km and extending to 1300 m.a.s.l. in 24 h. Over the following days, field observations highlighted a marked decline in the effusive activity, accompanied by a gradual up-slope migration of the active lava flow fronts and by a decrease in the intensity of the explosive activity from the uppermost portion of the eruptive fissure. Beginning in June and lasting until the end of July, a sudden increase in the effusive activity from the upper vents along the eruptive fissure occurred, causing the lava flow fronts to extend to lower elevations. This period was also associated with an increase in the explosive activity. Over the following month (August), the effusive activity progressively decreased, with only one brief recovery occurring in mid-September 2008. For all the months that followed, the eruptive activity remained relatively constant at low intensity. The eruption ended on 6 July 2009, after almost 14 months of continuous lava effusion.

5.1. Eruption thermal activity

Using the satellite module of LAV@HAZARD, we obtained a detailed chronology of the thermal activity spanning the entire eruption. The analysis of the time series of radiance maps and radiative power, derived from SEVIRI and MODIS images collected from May 2008 to July 2009, allowed detection of diverse thermal phases each of which could be related to different effusive processes:

1. *10 May 2008 (the paroxysmal event).* The sequence of images gathered by SEVIRI on 10 May permitted us to closely follow the opening paroxysmal episode. Volcanic thermal anomalies were detected during the time interval spanning 14:00 to 19:00 GMT. However, the period was characterized by a thick cloud cover and ash emission that led to an underestimation of remotely sensed radiance. Despite the cloudy weather conditions, this fountain was the most powerful of those recorded by SEVIRI sensor between 2007 and 2008 reaching a value, at 15:00 GMT, of ~15 GW (Fig. 3).
2. *13–15 May 2008 (the early effusive phase).* SEVIRI data recorded from 9:15 to 10:30 GMT on 13 May revealed emission of an ash plume (Fig. 4, yellow pixels) that persisted until the first hot spot related to eruptive activity (Fig. 4, red pixels). No eruptive phenomena are observable in the SEVIRI image acquired at 9:15 GMT. The next image acquired at 9:30 GMT detected the beginning of an ash plume moving NE. This was associated with a lava fountain from the northern part of the eruptive fissure. The first hot spot, due to renewed lava output, was detected at 10:30 GMT. Images from MODIS were also analyzed showing a first hot spot at 9:50 GMT (Fig. 5a). High temporal resolution of SEVIRI enabled a precise timing of the early phase of the eruption, but information from MODIS revealed the point of most intense activity, registering a maximum value of radiative power of about 16 GW on 15 May at 00:15 GMT (Figs. 5–6).
3. *16 May–7 June 2008 (the waning phase).* From 16 May, a decline in the SEVIRI-recorded radiative power was matched by a change in the structure of the thermal anomaly, with the eastern portion of

Fig. 2. Sketch map of the lava flow field of the 2008–2009 eruption at Mt Etna (see inset). The most distant lava fronts stagnated about 3 km from the nearest village, Milo.

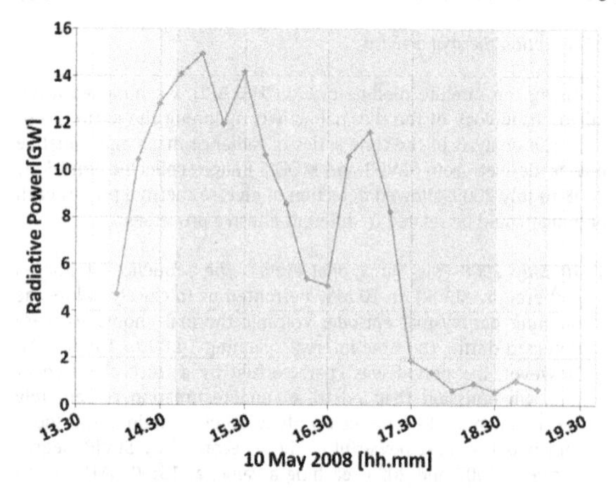

Fig. 3. Total radiative power recorded during the paroxysm occurred at Etna on 10 May 2008. The whole period was characterized by a thick cloud layer (yellow bar) and ash emission (orange bar).

the anomaly becoming less intense. The decrease in radiative power continued until 7 June, maintaining an average daily value of about 1–2 GW (Fig. 6). This is consistent with the low intensity of volcanic activity revealed by field observations during the same period.

4. *8 June–27 July 2008* (*the rising phase*). An increase in radiative power was recorded between 8 June and 27 July, reaching maximum values of about 10 GW on 21 June at 11:35 GMT by MODIS and 6 GW on 13 July at 08:15 GMT by SEVIRI (Fig. 6). This phase was characterized by high but scattered values of radiative power (such as the sudden decrease on 30 June); probably due to discontinuous ash emission during the Strombolian activity, which led to highly variable viewing conditions.
5. *28 July–13 September 2008* (*the falling phase*). A progressive decrease in radiative power was observed after 28 July, declining to a minimum value of less than 1 GW on 8 September (Fig. 6). The decreasing trend in the radiative power was interrupted on 11 September, when a peak value of 4 GW was recorded at 08:45 GMT by SEVIRI. This recovery of radiative power was accompanied by a weak explosive activity.
6. *14 September 2008–6 July 2009* (*the weak phase*). For all of the months that followed, the SEVIRI-derived radiative power plot showed constantly low values, maintaining a value of about 1 GW (Fig. 6). The last hot spot detected by SEVIRI occurred on 30 May 2009 at 05:00 GMT, with MODIS registering thermal anomalies until 26 June 2009 at 09:45 GMT.

Lava discharge rate estimates were computed (up to four times per hour) for the whole period of thermal emission from 13 May 2008 to 26 June 2009, converting the total radiative power measured from infrared satellite data (Fig. 7). We obtained minimum and maximum estimates for TADR by taking into account in the variability range of each lava parameter (Table 1) the largest and smallest

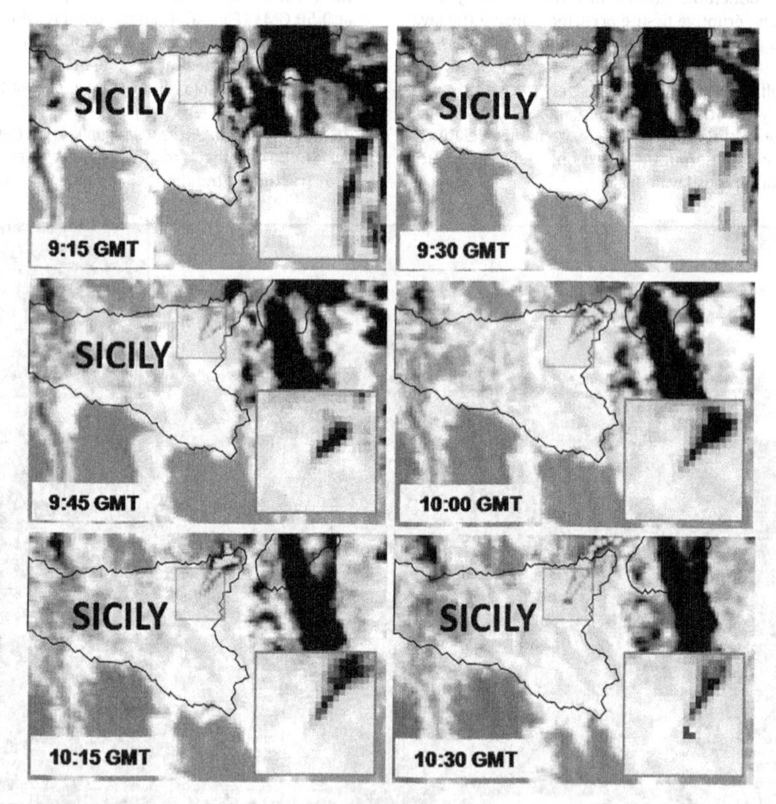

Fig. 4. A sequence of SEVIRI images recorded from 9:15 to 10:30 GMT on 13 May. At 9:15 GMT, no evident eruptive phenomena are observable. At 9:30 GMT the image shows the beginning of an ash plume (yellow pixels) moving toward NE, which was associated to the lava fountain from the northern part of the eruptive fissure. The first hot spot, due to the increased lava output and thermal anomaly, was detected by HOTSAT at 10:30 GMT.

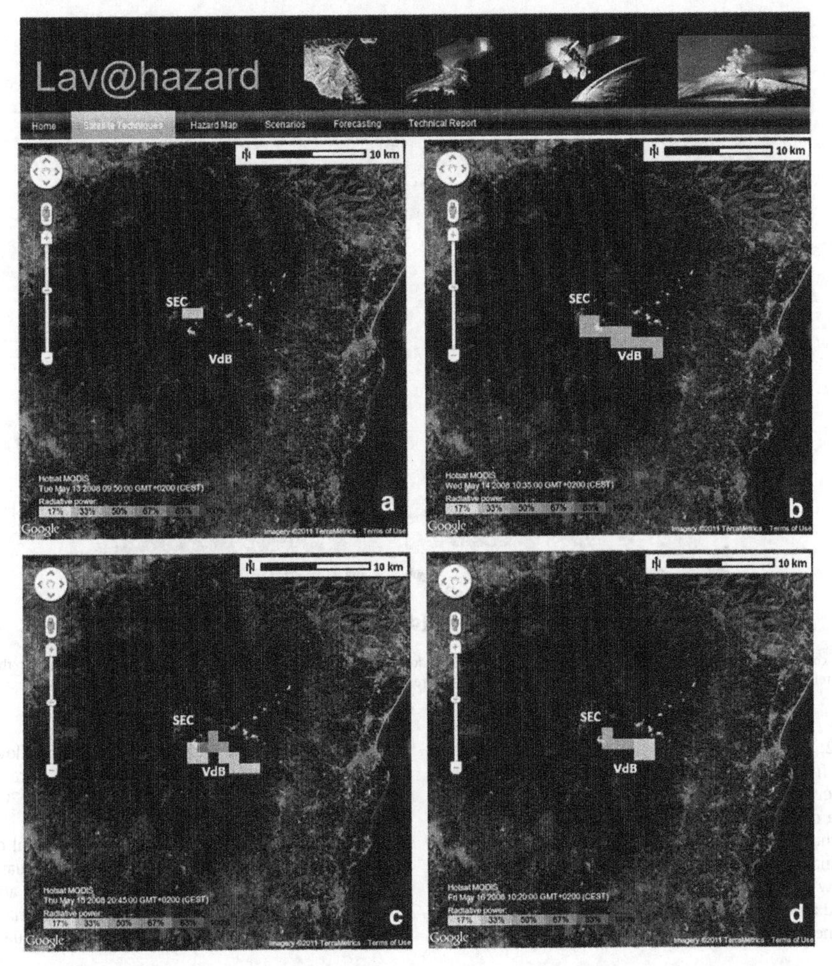

Fig. 5. Hotspot detected during 13–16 May 2008 by MODIS data. Four instants are reported on Google Map: a) 13 May at 09:50 GMT, b) 14 May at 10:35 GMT, c) 15 May at 20:45 GMT, d) 16 May at 10:20 GMT. Different colors are associated to different values of radiative power as reported in the bottom legend.

values, respectively. Peak estimates for TADR were reached on 15 May, ranging between 15 and 28 $m^3 s^{-1}$, and 21 June, ranging between 10 and 18 $m^3 s^{-1}$. By integrating minimum and maximum TADR values over the entire eruptive period preceding each successive measurement, we computed two cumulative curves of erupted lava (Fig. 7). Over the entire period of thermal activity (which ceased on 26 June 2009), we estimate a lava volume of between 32 and 61 million cubic meters was erupted. While erupted volumes by 8 June 2008 were between 3 and 6 million cubic meters, by 13 September they were between 23 and 43 million cubic meters. Showing that about 2/3 of the lava was erupted during the first 4 months of the eruption.

5.2. Lava flow simulations

By using the scenario forecasting module of LAVA@HAZARD, we calculated the lava flow paths using the estimates for TADR derived from thermal satellite data for the first 76 days of eruption, i.e. during the period between 13 May and 27 July 2008. This period of time corresponds to the first three phases of the effusive eruption (phases 2–4) during which time TADR steadily increased to a peak, and by the end of which the flow-field had attained 100% of its final length. We simulated the evolution of the flow field, step-by-step, over this period using, as our topographic base, a suitable DEM created by up-dating the 2005 2-m resolution DEM with data collected during an aerophotogrammetry flight of 2007 (Neri et al., 2008).

The actual flow areas and two simulated scenarios, obtained by using the minimum and maximum derived TADR, are shown in Fig. 8. The main differences between the simulated and observed lava flow paths are mainly governed by the values of TADR used and topography. In fact, the misfit between the actual and modeled length evolution occurs because the simulated flow is always slower than the actual flow, although the final distribution of lava-inundated areas is almost the same. This is probably due to differences between the maximum value of the satellite-derived lava discharge rate and the real value.

5.3. Discussion

The uncertainty in satellite derived TADR estimates is quite large, up to about 50%, but it is comparable to the error in field-based effusion rate measurements (Calvari et al., 2003; Harris & Baloga, 2009;

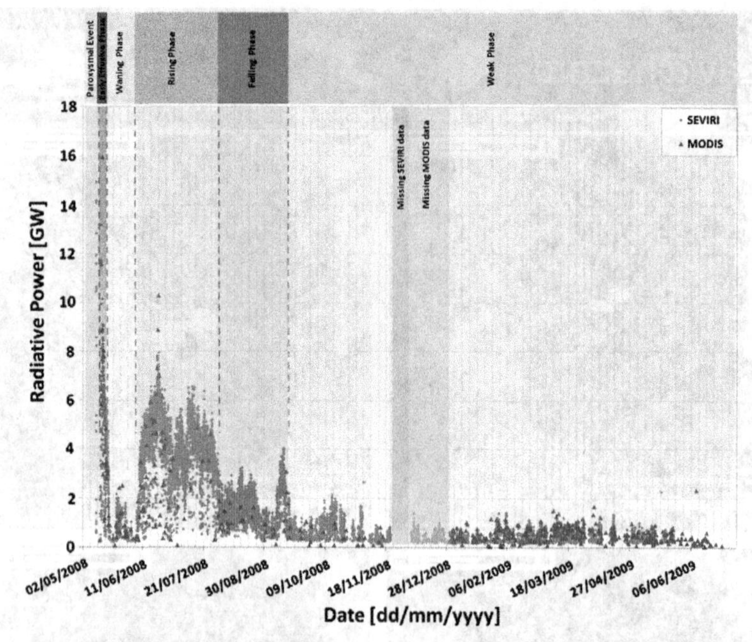

Fig. 6. Radiative power computed during the 2008–2009 Etna eruption by SEVIRI (blue dots) and MODIS (red triangles) data. Colored bars correspond to the six different thermal phases. Gray areas are related to missing data of MODIS (gray) and SEVIRI (dark gray) respectively.

Harris & Neri, 2002; Harris et al., 2007; Sutton et al., 2003). The main uncertainties arise from the lack of constraint on the lava parameters used to convert thermal flux to TADR (Harris & Baloga, 2009). Moreover, the presence of ash strongly affected the satellite-based estimates of lava discharge rate. This led to an underestimation of the satellite-derived final volume, and to a difference in the timing of simulated lava flow emplacement. Moreover, a certain discrepancy between actual and modeled flow areas is to be expected, as changes in the contemporaneous effusion rate take time to be translated into a change in active flow area (that is, the active flow area at any one time is a function of the antecedent effusion rate, rather than the instantaneous effusion rate) (Harris & Baloga, 2009; Wright et al., 2001).

Topographic effects also influence the spatial distribution of the flow. While the spatial overlap between the actual lava flow and the modeled flow driven by the minimum TADR is about 25%, the flow path simulated with the maximum TADR overlaps the actual lava flow by 60% (suggesting that the maximum value is more realistic).

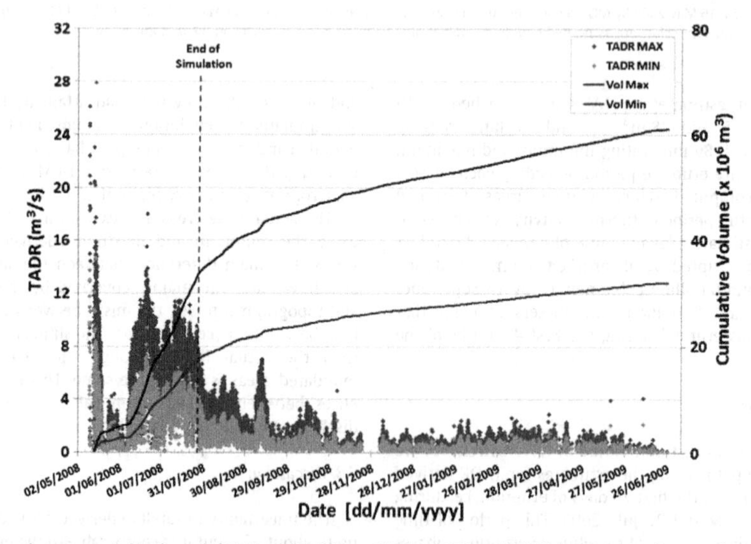

Fig. 7. TADRs and cumulative volumes computed in the period 13 May 2008–26 June 2009 by MODIS and SEVIRI data. Purple and green colors correspond to maximum and minimum values, respectively.

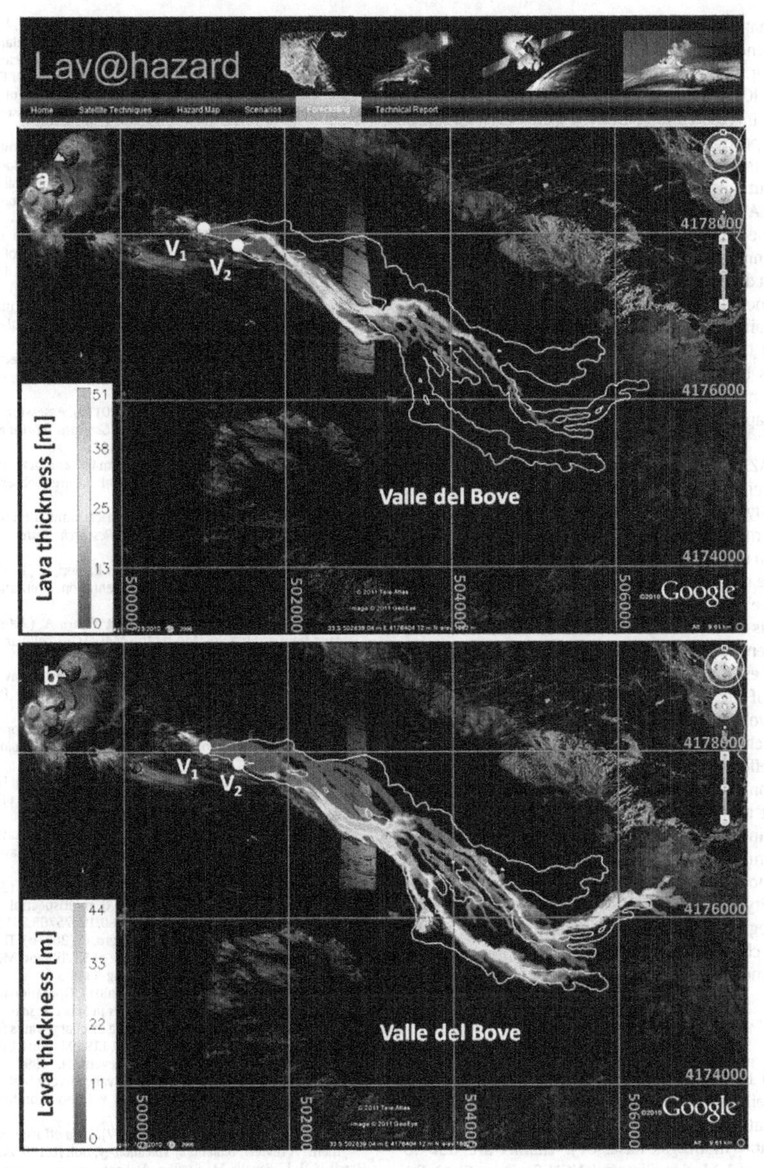

Fig. 8. Lava flow paths simulated by MAGFLOW (V_1 UTM coordinates: 500901E/4177991N, V_2 UTM coordinates: 501263E/4177852N) using the (a) minimum and the (b) maximum TADR. Simulations were performed from 13 May to 27 July 2008 and compared with the actual lava flow field (white contour; Behncke & Calvari, 2008). After this data no significant change appears in the final lava flow path.

The final simulated flow-fields are also always lacking in the northern part mapped for actual flow-field (Fig. 8). These space and time differences were probably due to the DEM, which does not include lava fountains emitted since 2007. These modified the morphology of the upper VdB so that our topography is out-of-date in this northern sector where we experience misfit. However, the zones covered by the simulated lava flow fit well with most of the observed lava flow field.

During the early phase of the eruption, several other simulations were performed to verify, which conditions, especially for the values of TADR, could be considered hazardous for the nearest infrastructures and villages. For all simulated scenarios, no lava flow path extended beyond the VdB area, even if higher values of TADR (increased of 50% of maximum value of satellite-derived TADR) were used as input of MAGFLOW. The most distant lava fronts always stagnated about 3 km from, Milo.

6. Concluding remarks

The LAV@HAZARD web-GIS framework shows great promise as a tool to allow tracking and prediction of effusive eruptions, allowing us to better assess lava flow hazards during a volcanic eruption in real-time. By using satellite-derived discharge rates to drive a lava flow emplacement model, LAV@HAZARD has the capability to forecast the lava flow hazards, allowing the estimation of the inundation area extent, the time required for the flow to reach a particular point, and the resulting morphological changes. We take advantage of the flexibility of the HOTSAT thermal monitoring system to process, in real time, satellite images coming from sensors with different spatial, temporal and spectral resolutions. In particular, HOTSAT was designed to ingest infrared satellite data acquired by the MODIS and SEVIRI sensors to output hot spot location, lava thermal flux and discharge rate. We use LAV@HAZARD to merge this output with the MAGFLOW physics-based model to simulate lava flow paths and to update, in a timely manner, flow simulations. In such a way, any significant changes in lava discharge rate are included in the predictions. MAGFLOW was implemented on the last generation of CUDA-enabled cards allowing us to gain a significant benefit in terms of computational speed thanks to the parallel nature of the hardware. All this useful information has been gathered into the LAV@HAZARD platform which, due to the high degree of interactivity, allows generation of easily readable maps and a fast way to explore alternative scenarios.

We tested LAV@HAZARD in an operational context as a support tool for decision makers using the 2008–2009 lava flow-forming eruption at Mt Etna. By using thermal infrared satellite data with low spatial and high temporal resolution, we continuously measured the temporal changes in thermal emission (at rates of up to four times per hour) for the entire duration of the eruption (also for short-lived phenomena such as the 10 May paroxysm preceding the eruption). The time-series analysis of radiative power, derived from SEVIRI and MODIS, was able to identify six different phases of eruptive activity. Satellite-derived TADR estimates were also computed and integrated for the whole period of the eruption (almost 14 months), showing that about 2/3 of lava volume was erupted during the first 4 months. These time-varying discharge rates were then used in the MAGFLOW model, allowing us to effectively simulate the advance time for active fronts and to produce on-the-fly eruptive scenarios fed by up-to-date TADRs. We found that no lava flows were capable of flowing over distances sufficient to invade vulnerable areas on the flanks of Etna. A comparison with actual, mapped, lava flow-field areas permitted validation of our methodology, predictions and results. Validations confirmed the reliability of LAV@HAZARD and its underlying data source and methodologies, as well as the potential of the whole integrated processing chain, as an effective strategy for real-time monitoring and assessment of volcanic hazard.

Acknowledgment

We are grateful to European Organisation for the Exploitation of Meteorological Satellites (EUMETSAT) for SEVIRI data (www.eumetsat.int) and to National Aeronautics and Space Administration (NASA) for MODIS data (modis.gsfc.nasa.gov). Thanks are due to Maria Teresa Pareschi and Marina Bisson (INGV-Sezione Pisa) for making the Digital Elevation Model of Etna available. This study was performed with the financial support from the V3-LAVA project (DPC-INGV 2007–2009 contract). This manuscript benefited greatly from the encouragement, comments, and suggestions of Andrew Harris, two anonymous reviewers, and the Editor Marvin Bauer.

References

Behncke, B., & Calvari, S. (2008). Etna: 6-km-long lava flow; ash emissions; 13 May 2008 opening of a new eruptive fissure. *Bull. Global Volcanism Network, 33*(5), 11–15.

Behncke, B., Neri, M., & Nagay, A. (2005). Lava flow hazard at Mount Etna (Italy): New data from a GIS-based study. In M. Manga, & G. Ventura (Eds.), *Kinematics and dynamics of lava flows. Spec. Pap. Geol. Soc. Am., 396.* (pp. 189–208), doi:10.1130/0-8137-2396-5.189(13).

Bilotta, G., Rustico, E., Hérault, A., Vicari, A., Del Negro, C., & Gallo, G. (2011). Porting and optimizing MAGFLOW on CUDA. *Annals of Geophysics, 54*, 5, doi:10.4401/ag-5341.

Bilotta, G., Cappello, A., Hérault, A., Vicari, A., Russo, G., & Del Negro, C. (Unpublished results). Sensitivity analysis of the MAGFLOW Cellular Automaton model, *Environmental Modelling and Software*, submitted for publication, July 2011.

Bonaccorso, A., Bonforte, A., Calvari, S., Del Negro, C., Di Grazia, G., Ganci, G., Neri, M., Vicari, A., & Boschi, E. (2011a). The initial phases of the 2008–2009 Mount Etna eruption: A multidisciplinary approach for hazard assessment. *Journal of Geophysical Research, 116*, B03203, doi:10.1029/2010JB007906.

Bonaccorso, A., Caltabiano, T., Currenti, G., Del Negro, C., Gambino, S., Ganci, G., Giammanco, S., Greco, F., Pistorio, A., Salerno, G., Spampinato, S., & Boschi, E. (2011b). Dynamics of a lava fountain revealed by geophysical, geochemical and thermal satellite measurements: The case of the 10 April 2011 Mt Etna eruption. *Geophysical Research Letters, 38*, doi:10.1029/2011GL049637.

Calvari, S., Neri, M., & Pinkerton, H. (2003). Effusion rate estimations during the 1999 summit eruption on Mount Etna, and growth of two distinct lava flow fields. *Journal of Volcanology and Geothermal Research, 119*, 107–123, doi:10.1016/S0377-0273(02)00308-6.

Calvari, S., & Pinkerton, H. (1998). Formation of lava tubes and extensive flow field during the 1991–93 eruption of Mount Etna. *Journal of Geophysical Research, 103*, 27291–27302.

Cappello, A., Vicari, A., & Del Negro, C. (2011a). Retrospective validation of a lava flow hazard map for Mount Etna volcano. *Annals of Geophysics, 54*, 5, doi:10.4401/ag-5345.

Cappello, A., Vicari, A., & Del Negro, C. (2011b). Assessment and modeling of lava flow hazard on Etna volcano. *Bollettino di Geofisica Teorica e Applicata, 52*, 299–308, doi:10.4430/bgta0003 n.2.

Carter, A., & Ramsey, M. (2010). Long-term volcanic activity at Shiveluch Volcano: Nine years of ASTER spaceborne thermal infrared observations. *Remote Sensing, 2*, 2571–2583, doi:10.3390/rs2112571.

Costa, A., & Macedonio, G. (2005). Numerical simulation of lava flows based on depth-averaged equations. *Geophysical Research Letters, 32*, L05304, doi:10.1029/2004GL021817.

Crisci, G. M., Di Gregorio, S., Pindaro, O., & Ranieri, G. (1986). Lava flow simulation by a discrete cellular model: first implementation. *International Journal of Modelling and Simulation, 6*, 137–140.

Del Negro, C., Fortuna, L., Herault, A., & Vicari, A. (2008). Simulations of the 2004 lava flow at Etna volcano by the MAGFLOW cellular automata model. *Bulletin of Volcanology*, doi:10.1007/s00445-007-0168-8.

Del Negro, C., Fortuna, L., & Vicari, A. (2005). Modelling lava flows by Cellular Nonlinear Networks (CNN): Preliminary results. *Nonlinear Processes in Geophysics, 12*, 505–513.

Dragoni, M., Bonafede, M., & Boschi, E. (1986). Downslope flow models of a Bingham liquid: Implications for lava flows. *Journal of Volcanology and Geothermal Research, 30*, 305–325.

Flynn, L. P., Mouginis-Mark, P. J., & Horton, K. A. (1994). Distribution of thermal areas on an active lava flow field: Landsat observations of Kilauea, Hawaii, July 1991. *Bulletin of Volcanology, 56*, 284–296.

Francis, P. W., & Rothery, D. A. (1987). Using the Landsat Thematic Mapper to detect and monitor active volcanoes: An example from Lascar volcano, Northern Chile. *Geology, 15*, 614–617.

Ganci, G., Vicari, A., Bonfiglio, S., Gallo, G., & Del Negro, C. (2011a). A texton-based cloud detection algorithm for MSG-SEVIRI multispectral images. *Geomatics, Natural Hazards and Risk, 2*, 1–12, doi:10.1080/19475705.2011.578263 3.

Ganci, G., Vicari, A., Fortuna, L., & Del Negro, C. (2011b). The HOTSAT volcano monitoring system based on a combined use of SEVIRI and MODIS multispectral data. *Annals of Geophysics, 54*, doi:10.4401/ag-5338 5.

Govaerts, Y., Arriaga, A., & Schmetz, J. (2001). Operational vicarious calibration of the MSG/SEVIRI solar channels. *Advances in Space Research, 28*, 21–30.

Harris, A. J. L., & Baloga, S. M. (2009). Lava discharge rates from satellite-measured heat flux. *Geophysical Research Letters, 36*, L19302, doi:10.1029/2009GL039717.

Harris, A. J. L., Blake, S., Rothery, D., & Stevens, N. (1997). A chronology of the 1991 to 1993 Mount Etna eruption using advanced very high resolution radiometer data: implications for real-time thermal volcano monitoring. *Journal of Geophysical Research, 102*, 7985–8003.

Harris, A. J. L., Dehn, J., & Calvari, S. (2007). Lava effusion rate definition and measurement: A review. *Bulletin of Volcanology*, doi:10.1007/s00445-007-0120-y.

Harris, A. J. L., Favalli, M., Steffke, A., Fornaciai, A., & Boschi, E. (2010). A relation between lava discharge rate, thermal insulation, and flow area set using lidar data. *Geophysical Research Letters, 37*, L20308, doi:10.1029/2010GL044683.

Harris, A. J. L., Flynn, L. P., Keszthelyi, L., Mouginis-Mark, P. J., Rowland, S. K., & Resing, J. A. (1998). Calculation of lava effusion rates from Landsat TM data. *Bulletin of Volcanology, 60*, 52–71.

Harris, A. J. L., Flynn, L. P., Rothery, D. A., Oppenheimer, C., & Sherman, S. B. (1999). Mass flux measurements at active lava lakes: Implications for magma recycling. *Journal of Geophysical Research, 104*, 7117–7136.

Harris, A. J. L., Murray, J. B., Aries, S. E., Davies, M. A., Flynn, L. P., Wooster, M. J., Wright, R., & Rothery, D. A. (2000). Effusion rate trends at Etna and Krafla and their implications for eruptive mechanisms. *Journal of Volcanology and Geothermal Research, 102*, 237–270.

Harris, A. J. L., & Neri, M. (2002). Volumetric observations during paroxysmal eruptions at Mount Etna: Pressurized drainage of a shallow chamber or pulsed supply? *Journal of Volcanology and Geothermal Research, 116*, 79–95, doi:10.1016/S0377-0273(02)00212-3.

Harris, A., & Rowland, S. (2001). FLOWGO: A kinematic thermorheological model for lava flowing in a channel. *Bulletin of Volcanology, 63*, 20–44.

Harris, A. J. L., & Rowland, S. K. (2009). Effusion rate controls on lava flow length and the role of heat loss: a review. In T. Thordarson, S. Self, G. Larsen, S. K. Rowland, & A.

Hoskuldsson (Eds.), *Studies in volcanology: The legacy of George Walker. Special Publications of IAVCEI, 2.* (pp. 33–51) London: Geological Society978-1-86239-280-9.

Harris, A. J. L., & Stevenson, D. S. (1997). Thermal observations of degassing open conduits and fumaroles at Stromboli and Vulcano using remotely sensed data. *Journal of Volcanology Geothermal Research, 76*, 175–198.

Harris, A. J. L., Vaughan, R. A., & Rothery, D. A. (1995). Volcano detection and monitoring using AVHRR data: The Krafla eruption,1984. *International Journal of Remote Sensing, 16*, 1001–1020.

Herault, A., Vicari, A., Ciraudo, A., & Del Negro, C. (2009). Forecasting lava flow hazard during the 2006 Etna eruption: Using the Magflow cellular automata model. *Computer & Geosciences, 35*(5), 1050–1060.

Hirn, B., Di Bartola, C., Laneve, G., Cadau, E., & Ferrucci, F. (2008). SEVIRI onboard Meteosat Second Generation, and the quantitative monitoring of effusive volcanoes in Europe and Africa. *Proceedings IGARSS 2008* (pp. 4–11). July 2008.

Hirn, B., Ferrucci, F., & Di Bartola, C. (2007). Near-tactical eruption rate monitoring of Pu'u O'o (Hawaii) 2000–2005 by synergetic merge of payloads ASTER and MODIS. *Geoscience and Remote Sensing Symposium, IGARSS 2007. IEEE International, 23–28 July 2007.* (pp. 3744–3747), doi:10.1109/IGARSS.2007.4423657.

Hulme, G. (1974). The interpretation of lava flow morphology. *Geophysical Journal of the Royal Astronomical Society, 39*, 361–383.

Kaneko, T., Wooster, M. J., & Nakada, S. (2002). Exogenous and endogenous growth of the Unzen lava dome examined by satellite infrared image analysis. *Journal of Volcanology and Geothermal Research, 116*, 151–160.

Miyamoto, H., & Sasaki, S. (1997). Simulating lava flows by an improved cellular automata method. *Computers & Geosciences, 23*, 283–292.

Neri, M., Mazzarini, F., Tarquini, S., Bisson, M., Isola, I., Behncke, B., & Pareschi, M. T. (2008). The changing face of Mount Etna's summit area documented with Lidar technology. *Geophysical Research Letters, 35*, L09305, doi:10.1029/2008GL033740.

Oppenheimer, C. (1991). Lava flow cooling estimated from Landsat Thematic Mapper infrared data: the Lonquimay eruption (Chile, 1989). *Journal of Geophysical Research, 96*, 21865–21878.

Oppenheimer, C., Francis, P. W., Rothery, D. A., Carlton, R. W. T., & Glaze, L. S. (1993). Infrared image analysis of volcanic thermal features: Láscar volcano, Chile, 1984–1992. *Journal of Geophysical Research, 98*, 4269–4286.

Pieri, D. C., & Abrams, M. J. (2005). ASTER observations of thermal anomalies preceding the April 2003 eruption of Chikurachki Volcano, Kurile Islands, Russia. *Remote Sensing of Environment, 99*, 84–94.

Pinkerton, H., & Wilson, L. (1994). Factors controlling the lengths of channel-fed lava flows. *Bulletin of Volcanology, 56*, 108–120.

Rose, W. I., & Mayberry, G. C. (2000). Use of GOES thermal infrared imagery for eruption scale measurements, Soufrière Hills, Montserrat. *Geophysical Research Letters, 27*, 3097–3100.

Running, S. W., Justice, C., Salomonson, V., Hall, D., Barker, J., Kaufman, Y., Strahler, A., Huete, A., Muller, J. P., Vanderbilt, V., Wan, Z. M., Teillet, P., & Carneggie, D. (1994). Terrestrial remote sensing science and algorithms planned for EOS/MODIS. *International Journal of Remote Sensing, 15*, 3587–3620.

Scifoni, S., Coltelli, M., Marsella, M., Proietti, C., Napoleoni, Q., Vicari, A., & Del Negro, C. (2010). Mitigation of lava flow invasion hazard through optimized barrier configuration aided by numerical simulation: The case of the 2001 Etna eruption. *Journal of Volcanology and Geothermal Research, 192*, 16–26, doi:10.1016/j.jvolgeores.2010.02.002.

Sutton, A. J., Elias, T., & Kauahikaua, J. (2003). Lava-effusion rates for the Pu'u'O'o-Kupaianaha eruption derived from SO2 emissions and very low frequency (VLF) measurements. *United States Geological Survey Professional Paper, 1676*, 137–148.

Vicari, A., Bilotta, G., Bonfiglio, S., Cappello, A., Ganci, G., Hérault, A., Rustico, E., Gallo, G., & Del Negro, C. (2011a). LAV@HAZARD: A web-GIS interface for volcanic hazard assessment. *Annals of Geophysics, 54*, 5, doi:10.4401/ag-5347.

Vicari, A., Ciraudo, A., Del Negro, C., Herault, A., & Fortuna, L. (2009). Lava flow simulations using discharge rates from thermal infrared satellite imagery during the 2006 Etna eruption. *Natural Hazards, 50*, 539–550, doi:10.1007/s11069-008-9306-7.

Vicari, A., Ganci, G., Behncke, B., Cappello, A., Neri, M., & Del Negro, C. (2011b). Near-real-time forecasting of lava flow hazards during the 12–13 January 2011 Etna eruption. *Geophysical Research Letters, 38*, doi:10.1029/2011GL047545.

Vicari, A., Herault, A., Del Negro, C., Coltelli, M., Marsella, M., & Proietti, C. (2007). Simulations of the 2001 lava flow at Etna volcano by the Magflow Cellular Automata model. *Environmental Modelling & Software, 22*, 1465–1471.

Walker, G. P. L. (1973). Lengths of lava flows. *Philosophical Transactions of the Royal Society of London, 274*, 107–118.

Wooster, M. J., Roberts, G., Perry, G. L. W., & Kaufman, Y. J. (2005). Retrieval of biomass combustion rates and totals from fire radiative power observations: FRP derivation and calibration relationships between biomass consumption and fire radiative energy release. *Geophysical Research Letters, 110*, D24311, doi: 10.1029/2005JD006318 1–24.

Wooster, M., Zhukov, B., & Oertel, D. (2003). Fire radiative energy release for quantitative study of biomass burning: derivation from the BIRD experimental satellite and comparison to MODIS fire products. *Remote Sensing of Environment, 86*, 83–107.

Wright, R., Blake, S., Harris, A., & Rothery, D. (2001). A simple explanation for the space-based calculation of lava eruption rates. *Earth and Planetary Science Letters, 192*, 223–233.

Wright, R., Flynn, L. P., Garbeil, H., Harris, A. J. L., & Pilger, E. (2004). MODVOLC: near-realtime thermal monitoring of global volcanism. *Journal of Volcanology and Geothermal Research, 135*, 29–49.

Wright, R., Garbeil, H., & Harris, A. J. L. (2008). Using infrared satellite data to drive a thermo-rheological/stochastic lava flow emplacement model: A method for near-real-time volcanic hazard assessment. *Geophysical Research Letters, 35*, L19307, doi:10.1029/2008GL035228.

Wright, R., Rothery, D. A., Blake, S., & Harris, A. J. L. (1999). Simulating the response of the EOS Terra ASTER sensor to high-temperature volcanic targets. *Geophysical Research Letters, 26*, 1773–1776.

Young, P., & Wadge, G. (1990). FLOWFRONT: simulation of a lava flow. *Computers & Geosciences, 16*, 1171–1191.

答案

任务 2.1 论文各部分使用的标题和小标题

Headings and subheadings for PEA1

Summary

Keywords

Introduction

Results

- Cloning of GmDmt1;1
- Gene expression
- Protein localisation
- Functional analysis in yeast

Discussion

- GmDmt1;1 can transport ferrous iron
- Specificity of GmDmt1;1
- Localisation and function of GmDmt1;1
- Regulation of GmDmt1;1 expression
- Conclusion

Experimental procedures

- Plant growth
- Isolation of GmDmt1;1
- Northern analysis
- Antibody generation and Western immunoblot analysis
- Symbiosome isolation and nodule membrane purification
- Functional expression in yeast

Acknowledgements

References

Headings and subheadings for PEA2

Summary

Keywords

Introduction

Methods

- Study system
- The invader

Field experiment

Statistical analysis

Model

Results

Discussion

Simulated urchin/mollusc disturbances

Propagule pressure and invasion success

Conclusions

Acknowledgements

References

Supplementary material

Headings and subheadings for PEA3

Abstract

1. Introduction

2. HOTSAT satellite monitoring system

2.1 Hot spot detection

2.2 Heat flux and TADR conversion

3. MAGFLOW lava flow simulator

4. LAV@HAZARD web-GIS framework

5. Case study：2008-2009 Etna eruption

5.1 Eruption thermal activity

5.2 Lava flow simulations

5.3 Discussion

6. Concluding remarks

Acknowledgment

References

任务 2.3 PEA 的结构

PEA1 属于 AIRDaM 结构。

PEA2 属于 AIMRaD 结构，在结尾处还有独立的结论部分。

PEA3 属于 AIBC 结构，正文部分包含四个主题，结论部分对应的标题是 Concluding remarks。

任务 2.4 预测

... yielded a total of... (R)

The aim of the work described... (I)

...was used to calculate... (M 或 R)

There have been few long-term studies of ... (I)

The vertical distribution of... was determined by ... (M 或 R)

This may be explained by ... (D)

Analysis was carried out using ... (M)

...was highly correlated with ... (R)

任务 3.1 审稿人在关注论文的哪个部分?

见表 AP1。

表 AP1 任务 3.1: 审稿人在关注论文的哪个部分?

Referee criterion	Likely location of evidence
1. Is the contribution new?	I (also stated in A, but no room to demonstrate it there)
2. Is the contribution significant?	I and D(also stated in A)
3. Is it suitable for publication in the Journal?	T,I,A
4. Is the organization acceptable?	All
5. Do the methods and the treatment of results conform to acceptable scientific standards?	M and R
6. Are all conclusions firmly based in the data presented?	R compared to D and A
7. Is the length of the paper satisfactory?	All
8. Are all illustrations required?	Photographs
9. Are all the figures and tables necessary?	Figures and tables
10. Are figure legends and table titles adequate?	As above
11. Do the title and Abstract clearly indicate the content of the paper?	T,A and all
12. Are the references up to date, complete, and the journal titles correctly abbreviated?	Ref
13. Is the paper excellent, good, or poor?	All

任务 3.2 论文标题信息提取

标题 A: *Use of in situ ^{15}N-labelling to estimate the total below-ground nitrogen of pasture legumes in intact soil-plant systems*

信息:

- The paper focuses on a particular method (*in situ* ^{15}N-labelling) and on results obtained using it.
- The parameter measured was total below-ground nitrogen.
- The measurement site/context was undisturbed systems involving both plants and soil.
- The plants used were pasture legumes.

问题(仅供参考):

- Why is this method suitable to measure this parameter in this context?
- Did the method provide reliable measurements?

- How was the accuracy of the measurements verified?
- How many legumes were studied and how did the results vary between them?
- What soil types were involved?
- Could this method be used for other plant/soil systems?

标题 B：*Short-and long-term effects of disturbance and propagule pressure on a biological invasion*

信息：

- The paper reports the effects of two factors (disturbance and propagule pressure) on one biological invasion.
- Results are reported over two time frames：short term and long term.
- The focus of the paper is on generalizations from the findings that apply to biological invasion in general (because no details are given in the title about the specific organisms or sites involved in this particular invasion)

问题（仅供参考）：

- What organisms and locations were involved in the invasion studied?
- What is the meaning of propagule pressure in this context?
- How are short term and long term defined in this paper?
- How do the specific results for this invasion provide evidence for the study of biological invasion in general?

标题 C：*The soybean NRAMP homologue，GmDMT1，is a symbiotic divalent metal transporter capable of ferrous iron transport*

信息：

- The paper reports the function (ability to transport divalent metals) of a newly identified entity which is an NRAMP homologue found in soybeans.
- The work reported in the paper shows that the homologue can transport one particular type of iron (ferrous iron) .
- The transport process is related to the symbiosis occurring in soybeans.

问题（仅供参考）：

- Why is the transport of ferrous iron significant in soybeans?
- How does the transport of divalent metals relate to the symbiosis?
- How was the function of this entity established?
- How does this finding contribute to the broader study of transporters?

标题 D：*An emergent strategy for volcano hazard assessment：From thermal satellite monitoring to lava flow modeling*

信息：

- The paper reports a method under development for evaluating the dangers presented by volcanoes.

- The work reported in the paper begins with thermal data obtained via satellite monitoring and presents a method for modeling the flow of lava.

问题（仅供参考）：

- What are the advantages of thermal satellite data for this type of application?
- Has the model been tested empirically or is this a theoretical paper?
- What aspects of the strategy are still to be developed?

任务 5.3　辨认图注说明的各个部分

见表 AP2、表 AP3 和表 AP4。

表 AP2　任务 5.3: PEA1

Sentence	Part
Number of *Sargassum muticum* (a) recruits and (b) adults in field experiment plots (900 cm^2).	Part 1
Propagule pressure is grams of reproductive tissue suspended over experimental plots at beginning of experiment.	Part 3
The average mass of an adult *S. muticum* (174g) is indicated by an arrow.	Part 5
Data are means ± 1 SE ($n=3$).	Part 4

表 AP3　任务 5.3: PEA2

Sentence	Part
Uptake of Fe(Ⅱ) by GmDmt1 in yeast.	Part 1
(a) Influx of $^{55}Fe^{2+}$ into yeast cells transformed with GmDmt1;1, *fet3fet4* cells	Part 1
were transformed with GmDmt1;1-pFL61 or pFL61 and then incubated with 1μM $^{55}FeCl_3$ (pH 5.5) for 5- and 10-min periods.	Part 3
Data presented are means ± SE of ^{55}Fe uptake between 5 and 10 min from three separate experiments (each performed in triplicate).	Part 4
(b) Concentration dependence of ^{55}Fe influx into *fet3fet4* cells transformed with GmDmt1;1-pFL61 or pFL61.	Part 1
Data presented are means ± SE of ^{55}Fe uptake over 5 min ($n=3$).	Part 4
The curve was obtained by direct fit to the Michaelis-Menten equation.	Part 2
Estimated K_M and V_{MAX} for GmDmt1;1 were(6.4 ± 1.1) μM Fe(Ⅲ) and (0.72 ± 0.08) nM Fe(Ⅲ)/min//mg protein, respectively.	Part 2
(c) Effect of other divalent cations on uptake of $^{55}Fe^{2+}$ into *fet3fet4* cells transformed with pFL61-GmDMT1;1.	Part 1
Data presented are means ± SE of ^{55}Fe (10μM) uptake over 10 min in the presence and absence of 100μM unlabeled Fe^{2+}, Cu^{2+}, Zn^{2+} and Mn^{2+}.	Parts 3 & 4

表 AP4　任务 5.3: PEA3

Sentence	Part
A sequence of SEVIRI images recorded from 9:15 to 10:30 GMT on 13 May.	Part 1
At 9:15 GMT, no evident eruptive phenomena are observable.	Part 3
At 9:30 GMT the image shows the beginning of an ash plume (yellow pixels) moving toward NE, which was associated to the lava fountain from the northern part of the eruptive fissure.	Part 3
The first hot spot, due to the increased lava output and thermal anomaly, was detected by HOTSAT at 10:30 GMT.	Part 3

任务 6.1 找出独立说明图表位置的语句

PEA1 无此类语句。

PEA2 只有一处独立成句，对图表位置进行说明的情况，而且是将“highlight + location”连成一句：“We plotted the proportion of plots in each treatment combination that were successfully invaded as a function of propagule pressure (Fig. 3).”这样做可能有两方面的原因：(1) 在句中可以使用主动语态，一些论文作者偏好这种直接的写作风格。(2) 整段话都是在试图回答作者在段首提出的问题 (How was the probability of successful invasion influenced by propagule pressure?)，这句话的写作风格需要与同一段落其他句子的风格保持一致。

PEA3 只有一处独立成句，对图表位置进行说明的情况，出现在第 5 部分“Case study”：“The actual flow areas and two simulated scenarios, obtained by using the minimum and maximum derived TADR, are shown in Fig. 8.”这句话没有强调研究发现，只是单纯说明了图表位置，而且作者使用被动语态，导致 Fig. 8 出现句尾，相对于图形所描绘的内容，Fig. 8 属于次要信息。

任务 7.1 “材料和方法”部分的呈现

见表 AP5。

表 AP5 任务 7.1:“材料和方法”部分的呈现

问题	PEA1	PEA2	PEA3
1. What subheadings are used in the section?	Methods; Study system; The invader; Field experiment; Statistical analysis; Model	Experimental procedures; Plant growth; Isolation of GmDmt1; 1; Northern analysis; Antibody generation and Western immunoblot analysis; Symbiosome isolation and nodule membrane purification	HOTSAT satellite monitoring system; Hot-spot detection; Heat flux and TADR conversion
2. i How do the subheadings relate to the end of the Introduction?	Very clear relation to the last paragraph of the Introduction. Wordings related to each subheading have been used there in describing the principal activity of the study and in almost the same order as the subheadings.	No specific relationship seen.	Specific relationship: HOTSAT was mentioned at the end of the Introduction as “recently developed” by the researchers and as a component of the new system presented in this paper.

续表

问题	PEA1	PEA2	PEA3
2. ii How do the subheadings relate to the subheadings in the Results section?	The last three subheadings come in the same order in which the Results are presented.	Results subheadings are not specifically related to Experimental procedure subheadings, but the order of the information in the Experimental procedure section follows closely the order in which the results are presented within that section.	No specific relationship seen.
3. Is the section easy for you to follow? Why? Or why not?	Aids to clarity include overview sentences at the start of paragraphs, before details are given.	Aids to clarity include frequent use of subheadings relating to order of information in Results and use of purpose phrases to show why steps were taken in relation to the experimental aims.	Aids to clarity include frequent use of introductory phrases with " to + verb " to show why actions were taken.

任务 7.3 主动句和被动句

我们在每篇 PEA 中都选取了两个句子，请与你的转化结果进行对比。

见表 AP6。

表 AP6 任务 7.3: 主动句和被动句

论文	原句	转化结果
PEA1	Soybean seeds were inoculated at planting with *Bradyrhizobium japonicum* USDA 110 … [passive]	We inoculated soybean seeds at planting with *Bradyrhizobium japonicum* USDA 110 … [active]
	Subsequent PCR experiments identified a full-length 1849-bp cDNA … [active]	A full-length 1849-bp cDNA was identified in subsequent PCR experiments … [passive]
PEA2	Control plots were not altered in any way, … [passive]	We did not alter control plots in any way, … [active]
	Each holdfast produces as many as 18 laterals in the early spring, … [active]	As many as 18 laterals are produced by each holdfast in early spring, … [passive]
PEA3	We integrated the HOTSAT system and the MAGFLOW model using a web-based Geographical Information System (GIS) framework … [active]	The HOTSAT system and the MAGFLOW model were integrated using … [passive]
	The satellite-derived discharge rate estimates are then used in the scenario forecasting module … [passive]	We then use the satellite-derived discharge rate estimates in the scenario forecasting module … [active]①

① 转化成主动句后句意有所改变。实际上，使用“estimates”的主体不是研究者，而是软件。这也说明，在特定语境下，被动语态是最佳选择。

任务 7.4 头重脚轻的被动句

改写后：

The soil water balance equation（Xin，1986；Zhu and Niu，1987）was used to compute actual evapotranspiration（T）for each crop，defined as the amount of precipitation for the period between sowing and harvesting the particular crop plus or minus the change in soil water storage in the 2m soil profile.

或改写成：

Actual evapotranspiration（T）for each crop was computed by the soil water balance equation（Xin，1986；Zhu and Niu，1987）. This measure is defined as the amount of precipitation for the period between sowing and harvesting the particular crop plus or minus the change in soil water storage in the 2m soil profile.

任务 8.1 引言部分的层级

见表 AP7、表 AP8 和表 AP9（见下页）。

任务 8.2 分析引言部分的层级 1

见表 AP10。

表 AP10 任务 8.2: 分析引言部分的层级 1

问题	PEA 1	PEA 2	PEA 3
是否有使用一般现在时的句子？有几句？	有，8 句	有，2 句	有，3 句
是否有使用现在完成时的句子？有几句？	没有	有，3 句	有，2 句
哪种时态使用更多？可能的原因是什么？	现在时更多 主要内容是在解释生物过程	完成时更多 主要内容是介绍研究领域的发展状况，迄今为止他人完成的工作	现在时更多 主要内容是说明研究地点的特征，在这里开展研究为何重要
有几句话使用了参考文献？①	1(共 8 句)	3(共 5 句)	3(共 5 句)
哪类句子没有引用参考文献？	概述读者们已经广泛接受的事实的句子	概述当前知识状况的句子	陈述事实和广泛接受的观点的句子

① 译者注：此处答案有误，本书作者在计数时只考虑了任务 8.1 中标示出 Stage 1 的一小部分文字（表 AP7、表 AP8 和表 AP9），层级 1 中包含参考文献的句子数量实际上大于答案给出的数字。

表 AP7 任务 8.1： PEA 2 引言部分的层级

引言部分	层级
Biological invasions are a global problem with substantial economic(Pimentel *et al*. 2005)and ecological(Mack *et al*. 2000)costs. Research on invasions has provided important insights into the establishment, spread and impact of non-native species. One key goal of invasion biology has been to identify the factors that determine whether an invasion will be successful(Williamson 1996). Accordingly, ecologists have identified several individual factors(e. g. disturbance and propagule pressure) that appear to exert strong controlling influences on the invasion process. However, understanding how these processes interact to regulate invasions remains a major challenge in ecology(D'Antonio *et al*. 2001; Lockwood *et al*. 2005; Von Holle & Simberloff 2005).	Stage 1 Stage 3(broad research niche, claiming significance)
Propagule pressure is widely recognized as an important factor that influences invasion success(MacDonald *et al*. 1989; Simberloff 1989; Williamson 1996; Lonsdale 1999; Cassey *et al*. 2005). Previous studies suggest that the probability of a successful invasion increases with the number of propagules released(Panetta & Randall 1994; Williamson 1989; Grevstad 1999), with the number of introduction attempts(Veltman *et al*. 1996), with introduction rate(Drake*et al*. 2005), and with proximity to existing populations of invaders(Bossenbroek *et al*. 2001). Moreover, propagule pressure may influence invasion dynamics after establishment by affecting the capacity of non-native species to adapt to their new environment(Ahlroth *et al*. 2003; Travis *et al*. 2005). Despite its acknowledged importance, propagule pressure has rarely been manipulated experimentally and the interaction of propagule pressure with other processes that regulate invasion success is not well understood(D'Antonio *et al*. 2001; Lockwood *et al*. 2005).	Stage 2 Stage 3(one component of the study, as indicated in the title)
Resource availability is a second key factor known to influence invasion success and processes that increase or decrease resource availability therefore have strong effects on invasions(Davis*et al*. 2000). Resource pre-emption by native species generates biotic resistance to invasion(Stachowicz *et al*. 1999; Naeem *et al*. 2000; Levine *et al*. 2004). Consequently, physical disturbance can facilitate invasions by reducing competition for limiting resources(Richardson & Bond 1991; Hobbs & Huenneke 1992; Kotanen 1997; Prieur-Richard & Lavorel 2000). In most communities disturbances occur via multiple mechanisms and the disturbances created by different agents vary in their intensity and frequency(D'Antonio *et al*. 1999). Recent empirical(Larson 2003; Hill *et al*. 2005)and theoretical(Higgins & Richardson 1998)studies suggest that not all types of disturbance have equivalent effects on the invasion process. Moreover, most of what we know about the effects of disturbance on invasions comes from short-term experimental studies. It is presently unclear how different disturbance agents influence long-term patterns of invasion.	Stage 2 Stage 3(another component of the study, as highlighted in the title)
In order for any invasion to be successful, propagule arrival must coincide with the availability of resources needed by the invading species(Davis *et al*. 2000). Therefore, the interaction between propagule pressure and processes that influence resource availability will ultimately determine invasion success(Brown & Peet 2003; Lockwood *et al*. 2005; Buckley *et al*. 2007). In this study we used the invasion of shallow, subtidal kelp communities in Washington State by the Japanese seaweed *Sargassum muticum* as a study system to better understand the effects of propagule pressure and disturbance on invasion. In a factorial field experiment we manipulated both propagule pressure and disturbance in order to examine how these factors independently and interactively influence *S. muticum* establishment in the short term. We supplement the experimental results with a parameterized integrodifference equation model, which we use to examine how different natural disturbance agents influence the spread of *S. muticum* through the habitat in the longer term. Although a successful invasion clearly requires both establishment and spread of the invader, most studies have looked at just one of these processes(Melbourne *et al*. 2007). We take an integrative approach by employing both a short-term experiment and a longer-term model, allowing us to examine the effects of disturbance and propagule limitation on the entire invasion process.	Stage 2 Stage 4(principal activities of the present study) Stage 5(value of the present study, claiming significance)

表 AP8 任务 8.1：PEA 1 引言部分的层级

引言部分	层级
Legumes form symbiotic associations with N_2-fixing soil-borne bacteria of the *Rhizobium* family. The symbiosis begins when compatible bacteria invade legume root hairs, signalling the division of inner cortical root cells and the formation of a nodule. Invading bacteria migrate to the developing nodule by way of an 'infection thread', comprised of an invaginated cell wall. In the inner cortex, bacteria are released into the cell cytosol, enveloped in a modified plasma membrane (the peribacteroid membrane (PBM)), to form an organelle-like structure called the symbiosome, which consists of bacteroid, PBM and the intervening peribacteroid space (PBS; Whitehead and Day, 1997). The bacteria, subsequently, differentiate into the N_2-fixing bacteroid form. The symbiosis allows the access of legumes to atmospheric N_2, which is reduced to NH_4^+ by the bacteroid enzyme nitrogenase. In exchange for reduced N, the plant provides carbon to the nodules to support bacterial respiration, a low-oxygen environment in the nodule suitable for bacteroid nitrogenase activity, and all the essential nutritional elements necessary for bacteroid activity.	Stage 1 (providing a context for the problem to be investigated)
Consequently, nutrient transport across the PBM is an important control mechanism in the promotion and regulation of the symbiosis.	Stage 3 (broad research niche, claiming importance)
Micronutrients such as iron are essential for bacteroid activity and nodule development. The demand for iron increases during symbiosis (Tang et al., 1990), where the metal is used for the synthesis of various iron-containing proteins in both the plant and the bacteroids. In the plant fraction, iron is an important part of the heme moiety of leghemoglobin, which facilitates the diffusion of O_2 to the symbiosomes in the affected cell cytosol (Appleby, 1984). In bacteroids, there are many iron-containing proteins involved in N_2 fixation, including nitrogenase itself and cytochromes used in the bacteroid electron-transport chain. In the soil, iron is often poorly available to plants as it is usually in its oxidised form Fe(Ⅲ), which is highly insoluble at neutral and basic pH. To compensate this, plants have developed two general strategies to gain access to iron from their localised environment. Strategy I involves secretion of phytosiderophores that aid in the solubilisation and uptake of Fe(Ⅲ), while Strategy Ⅱ involves initial reduction of Fe(Ⅲ) to Fe(Ⅱ) by a plasma membrane Fe(Ⅲ)-chelate reductase, followed by uptake of Fe(Ⅱ) (Romheld, 1987).	Stage 1 (another aspect of the context, iron, as indicated in the title)
The mechanism(s) involved in bacteroid iron acquisition within the nodule have been investigated at the biochemical level, and three activities have been identified (Day et al., 2001). Fe(Ⅲ) is transported across the PBM complexed with organic acids such as citrate, and accumulates in the PBS (Levier et al., 1996; Moreau et al., 1995), where it becomes bound to siderophore-like compounds (Wittenberg et al., 1996). Fe(Ⅲ) chelate reductase activity has been measured on isolated PBM, and Fe(Ⅲ) uptake into isolated symbiosomes is stimulated by Nicotinamide Adenine Dinucleotide (NADH), reduced form (Levier et al., 1996). However, Fe(Ⅱ) is also readily transported across the PBM and has been found to be the favoured form of iron taken up by bacteroids (Moreau et al., 1998).	Stage 2 (aspects of the problem already investigated by others)
The proteins involved in this transport have not yet been identified.	Stage 3 (more specific research gap)
Two classes of putative Fe(Ⅱ)-transport proteins (Irt/Zip and Dmt/Nramp) have been identified in plants (Belouchi et al., 1997; Curie et al., 2000; Eide et al., 1996;	Stage 2

续表

引言部分	层级
Thomine et al. ,2000). The Irt/Zip family was first identified in *Arabidopsis* by functional complementation of the yeast Fe(II) transport mutant DEY1453(*fet3fet4*; Eide et al. ,1996). *AtIrtl* expression is enhanced in roots when grown on low iron(Eide et al. ,1996), and appears to be the main avenue for iron acquisition in *Arabidopsis*(Vert et al. ,2002). Recently, a soybean Irt/Zip isologue, GmZip1, was identified and localised to the PBM in nodules(Moreau et al. ,2002). GmZip1 has been characterised as a symbiotic zinc transporter, which does not transport Fe(II). The second class of iron-transport proteins consists of the Dmt/Nramp family of membrane transporters, which were first identified in mammals as a putative defense mechanism utilised by macrophages against mycobacterium infection(Supek et al. ,1996; Vidal and Gros, 1994). Mutations in Nramp proteins in different organisms result in varied phenotypes including altered taste patterns in *Drosophila*(Rodrigues et al. ,1995), microcytic anaemia(mk) in mice and Belgrade rats(Fleming *et al.* ,1997) and loss of ethylene sensitivity in plants(Alonso *et al.* ,1999). The rat and yeast NRAMP homologues(DCT1 and SMF1, respectively) have been expressed in *Xenopus* oocytes and shown to be broad-specificity metal ion transporters capable of Fe(II), amongst other divalent cations, transport(Chen et al. ,1999; Gunshin et al. ,1997). The plant homologue, AtNramp1, complements the growth defect of the yeast Fe(II) transport mutant DEY1453, while other *Arabidopsis* members do not(Curie et al. ,2000; Thomine et al. ,2000). Interestingly, AtNramp1 overexpression in *Arabidopsis* also confers tolerance to toxic concentrations of external Fe(II)(Curie et al. ,2000), suggesting, perhaps, that it is localised intracellularly.	Stage 3(building the gap) Stage 2
In this study, we have identified a soybean homologue of the Nramp family of membrane proteins, GmDmt1;1. We show that GmDmt1;1 is a symbiotically enhanced plant protein, expressed in soybean nodules at the onset of nitrogen fixation, and is localised to the PBM. GmDmt1;1 is capable of Fe(II) transport when expressed in yeast. Together, the localisation and demonstrated activity of GmDmt1;1 in soybean nodules suggests that the protein is involved in Fe(II) transport and iron homeostasis in the nodule to support symbiotic N_2 fixation.	Stage 4(principal activity of the study and conclusion reached)

表 AP9 任务 8.1： PEA 3 引言部分的层级

引言部分	层级
Etna volcano, Italy, is characterized by persistent activity, consisting of degassing and explosive phenomena at its summit craters, as well as frequent flank eruptions. All eruption typologies can give rise to lava flows, which are the greatest hazard presented by the volcano to inhabited and cultivated areas(Behncke et al. ,2005).	Stage 1
The frequent eruptions of Mt Etna and hazard they pose represent an excellent opportunity to apply both spaceborne remote sensing techniques for thermal volcano	

续表

引言部分	层级
monitoring, and numerical simulations for predicting the area most likely to be inundated by lava during a volcanic eruption. Indeed, Etna has witnessed many tests involving the application of space-based remote sensing data to detect, measure and track the thermal expression of volcanic effusive phenomena(e. g. Ganci et al., 2011b; Harris et al., 1998; Wright et al., 2004), as well as a number of lava flow emplacement models(e. g. Crisci et al., 1986; Harris & Rowland, 2001; Vicari et al., 2007). These efforts have allowed key at-risk areas to be rapidly and appropriately identified(Ganci et al., 2011b; Wright et al., 2008). Over the last 25 years, satellite measurements in the thermal infrared have proved well suited to detection of volcanic thermal phenomena(e. g. Francis & Rothery, 1987; Harris et al., 1995) and to map the total thermal flux from active lava flows(e. g. Flynn et al., 1994; Harris et al., 1998). High spatial resolution data collected by Landsat and ASTER have been employed for the thermal analysis of active lava flows(Hirn et al., 2007; Oppenheimer, 1991), lava domes(Carter & Ramsey, 2010; Kaneko et al., 2002; Oppenheimer et al., 1993), lava lakes(Harris et al., 1999; Wright et al., 1999), and fumarole fields(Harris & Stevenson, 1997; Pieri & Abrams, 2005). Lower spatial, but higher temporal, resolution sensors, such the Advanced Very High Resolution Radiometer(AVHRR) and the MODerate Resolution Imaging Spectro-radiometer(MODIS), have also been widely used for infrared remote sensing of volcanic thermal features, as has GOES(e. g. Rose & Mayberry, 2000). The high temporal resolution(15 min) offered by the Spinning Enhanced Visible and Infrared Imager(SEVIRI), already employed for the thermal monitoring of effusive volcanoes in Europe and Africa(Hirn et al., 2008), has recently been exploited to estimate lava discharge rates for eruptive events of short duration(a few hours) at Mt Etna(Bonaccorso et al., 2011a, 2011b). These data have also proved capable of detecting, measuring and tracking volcanic thermal phenomena, despite the fact that the volcanic thermal phenomena are usually much smaller than the nominal pixel sizes of 1-4 km(Ganci et al., 2011a, 2011b; Vicari et al., 2011b). Great advances have also been made in understanding the physical processes that control lava flow emplacement, resulting in the development of a range of tools allowing the assessment of lava flow hazard. Many methods have been developed to simulate and predict lava flow paths and run-out distance based on various simplifications of the governing physical equations and on analytical and empirical modeling(e. g. : Hulme, 1974; Crisci et al., 1986; Young & Wadge, 1990; Miyamoto & Sasaki, 1997; Harris & Rowland, 2001; Costa & Macedonio, 2005; Del Negro et al., 2005). In its most simple form, such forecasting may involve the application of volcano-specific empirical length/effusion rate relationships to estimate flow length(e. g. Calvari & Pinkerton, 1998). At their most complex, such predictions may involve iteratively solving a system of equations that characterize the effects that cooling-induced changes in rheology have on the ability of lava to flow downhill, while taking into account spatial variations in slope determined from a digital elevation model(e. g. Del Negro et al., 2008). On Etna, the MAGFLOW Cellular Automata model has successfully been used to reproduce lava flow paths during the 2001, 2004 and 2006 effusive eruptions(Del	Stage 2

续表

引言部分	层级
Negro et al. ,2008;Herault et al. ,2009;Vicari et al. ,2007). More recently,it has been applied to consider the impact of hypothetical protective barrier placement on lava flow diversion(Scifoni et al. ,2010)and has been used for the production of a hazard map for lava flow invasion on Etna(Cappello et al. ,2011a,2011b). MAGFLOW is based on a physical model for the thermal and rheological evolution of the flowing lava. To determine how far lava will flow,MAGFLOW requires constraint of many parameters.	
However,after chemical composition,the instantaneous lava output at the vent is the principal parameter controlling the final dimensions of a lava flow(e. g. Harris & Rowland,2009;Pinkerton & Wilson,1994;Walker,1973). As such,any simulation technique that aims to provide reliable lava flow hazard assessments should incorporate temporal changes in discharge rate into its predictions in a timely manner.	Stage 3:need/problem to be addressed
Satellite remote sensing provides a means to estimate this important parameter in real-time during an eruption(e. g. Harris et al. ,1997;Wright et al. ,2001). A few attempts to use satellite-derived discharge rates to drive numerical simulations have already been made(e. g. Herault et al. ,2009;Vicari et al. ,2009;Wright et al. ,2008). These modeling efforts all used infrared satellite data acquired by AVHRR and/or MODIS to estimate the time-averaged discharge rate(TADR)following the methodology of Harris et al. (1997). Recently,we developed the HOTSAT multiplatform system for satellite infrared data analysis,which is capable of managing multispectral data from different sensors as MODIS and SEVIRI to detect volcanic hot spots and output their associated radiative power(Ganci et al. ,2011b). Satellite-derived output for lava flow modeling purposes has already been tested for the 12 - 13 January 2011 paroxysmal episode at Mt Etna by Vicari et al. , 2011b.	Stage 1/2-stage 3 implicit in "a few"
Here we present a new web-based GIS framework,named LAV@HAZARD,which integrates the HOTSAT system for satellite-derived discharge rate estimates with the MAGFLOW model to simulate lava flow paths. As a result,LAV@HAZARD now represents the central part of an operational monitoring system that allows us to map the probable evolution of lava flow-fields while the eruption is ongoing. Here we describe and demonstrate the operation of this LAV@HAZARD using a retrospective analysis of Etna's 2008—2009 flank eruption.	Stage 4:main activity of the paper
This eruption was exceptionally well documented by a variety of monitoring techniques maintained by INGV-CT including spaceborne thermal infrared measurements.	Stage 2
Within this operational role,HOTSAT is first used to identify thermal anomalies due to active lava flows and to compute the TADR. This is then used to drive MAGFLOW simulations,allowing us to effectively simulate the advance rate and maximum length for the active lava flows.	Stage 4
Because satellite-derived TADRs can be obtained in real times and simulations spanning several days of eruption can be calculated in a few minutes,such a combined approach has the potential to provide timely predictions of the areas likely to be inundated with lava,which can be updated in response to changes of the eruption conditions as detected by the current image conditions. If SEVIRI data are used,simulations can be updated every 15 min. Our results thus demonstrate how LAV@HAZARD can be exploited to produce realistic lava flow hazard scenarios and for helping local authorities in making decisions during a volcanic eruption.	Stage 5

任务 8.3 层级 1：从"国家"到"城市"

PEA1：

What is the "country"? Legume symbiotic associations.

The "province"? The peribacteroid membrane (PBM) and its role.

The "city"? Nutrient transport across the PBM.

PEA2：

What is the "country"? Biological invasions.

The "province"? Factors controlling the invasion process.

The "city"? The interaction of the factors and processes.

PEA3：

What is the "country"? Etna volcano.

The "province"? Hazard caused by lava flow from volcanoes.

The "city"? Applications of relevant technologies for hazard identification.

任务 8.4 辨识旧的信息

旧的信息已用下划线标出。

Legumes form symbiotic associations with N_2-fixing soil-borne bacteria of the *Rhizobium* family. The symbiosis begins when compatible bacteria invade legume root hairs, signalling the division of inner cortical root cells and the formation of a nodule. Invading bacteria migrate to the developing nodule by way of an 'infection thread', comprised of an invaginated cell wall. In the inner cortex, bacteria are released into the cell cytosol, enveloped in a modified plasma membrane (the peribacteroid membrane (PBM)), to form an organelle-like structure called the symbiosome, which consists of bacteroid, PBM and the intervening peribacteroid space (PBS; Whitehead and Day, 1997). The bacteria, subsequently, differentiate into the N_2-fixing bacteroid form. The symbiosis allows the access of legumes to atmospheric N_2, which is reduced to NH_4^+ by the bacteroid enzyme nitrogenase. In exchange for reduced N, the plant provides carbon to the nodules to support bacterial respiration, a low-oxygen environment in the nodule suitable for bacteroid nitrogenase activity, and all the essential nutritional elements necessary for bacteroid activity. Consequently, nutrient transport across the PBM is an important control mechanism in the promotion and regulation of the symbiosis.

任务 8.6 剽窃的鉴别

见表 AP11。

表 AP11 任务 8.6: 剽窃的鉴别

Version 2 涉嫌剽窃的句子	理由
However, this technique is not adaptable to all plants, particularly pasture species.	This sounds like the idea of the writer of the paragraph, but we know from Version 1 that it was originally the idea of Russell and Fillery (1996). Because there is no grammatical link between the two sentences, the reference in the first sentence does not apply to the second sentence. Note in Version 1 that the authors used both a grammatical link (they) and a tense marker (past tense *was not adaptable*) to indicate that the idea came from the cited work.

任务 8.7 指出研究空白:信号词

见表 AP12。

表 AP12 任务 8.7: 指出研究空白

McNeill et al. (1997)	PEA1	PEA2	PEA3
scarce, with little account taken of, is accordingly required, but, however	consequently, however, have not yet been identified, putative, appears to be	remains a major challenge, despite its acknowledged importance, rarely, is not well understood, it is presently unclear how, to better understand	should incorporate, in a timely manner

任务 8.9 层级 4 使用的句子模板

McNeill et al. (1997)

The experiments reported here were designed (i) to assess the use of [NP1] to [verb phrase], and (ii) to obtain quantitative estimates of [NP2].

PEA1

In this study we have identified [NP1], [NP2]. We show that [NP2] is [NP3], expressed in [NP4] at the onset of [NP5], and is localised to [NP6].

PEA2

In this study we used [NP1] as a study system to better understand the effects of [NP2] and [NP3] on [NP4]. In a [adjective] experiment we manipulated both [NP2] and [NP3] in order to examine how these factors [adverbs] influence [NP5] in the short term. We supplement the experimental results with [NP6], which we use to examine how different

[NP7] influence [NP8] in the longer term.

PEA3

Here we present [NP1], which integrates [NP2] with [NP3] to simulate [NP4] . ... Here we describe and demonstrate the operation of this [NP1] using a [adjective] analysis of [NP5] .

任务 8. 11 分析段落的主题句

见表 AP13。

表 AP13 任务 8. 11: 分析段落的主题句

来源	主题句	上文	下文
PEA1	Propagule pressure is widely recognized as an important factor that influences invasion success(references).	Refers to *propagule pressure* as one of two examples of factors influencing invasions.	Gives details of results of previous studies showing ways in which *propagule pressure* affects invasion success.
PEA2	Two classes of putative Fe(II)-transport proteins (Irt/Zip and Dmt/Nramp) have been identified in plants(references).	Ends by stating that the proteins involved are *unknown*, which links directly to *putative* (= possible candidates) in this sentence.	Gives details of research results on each of the two classes, in the same order in which they are referred to in the topic sentence (Irt/Zip and then Dmt/Nramp).
PEA 3	On Etna, the MAGFLOW Cellular Automata model has successfully been used to reproduce lava flow paths during the 2001, 2004 and 2006 effusive eruptions(references).	Describes many methods and models used to simulate and predict lava flows.	Reports several successful uses of the MAGFLOW model and introduces the constraints to the use of this and other models that are to be addressed in the present study.

任务 8. 12 新旧信息的编排

划线部分需要调整：

Pleuropneumonia (APP) surfaced in the Australian pig population during the first half of the 1980s and ten years later was regarded as one of the most costly and devastating diseases affecting the Australian pig industry. It can present as a dramatic clinical disease or as a chronic, production limiting disease in pig herds. A sudden increase in the number of sick and coughing pigs and a sharp rise in mortalities among grower/finisher pigs may herald an outbreak of APP in a herd. On the other hand, signs may be limited to a drop in growth rate and an increase in grade two pleurisy lesions in slaughter pigs.

可以调整语序为：

An outbreak of APP in a herd may be heralded by a sudden increase in the number of sick and coughing pigs and a sharp rise in mortalities among grower/finisher pigs.

注意：修改的实质就是将主动（may herald）改为被动（may be heralded）。在主动和被动间切换是非常有用的思维方式，它允许我们调整句中信息的呈现顺序。对于论文作者而言，如果想要满足新旧信息的编排规则，可以尝试使用这种思路对句子进行修改。

任务 8.13 改写头重脚轻的句子

1. 原句：In this project the *Rhizoctonia* populations of two field soils in the Adelaide Plains region of South Australia were characterised.

改写后：This project characterised the *Rhizoctonia* populations of two field soils in the Adelaide Plains region of South Australia. 还可以改写成：

The aim of this project was to characterise the *Rhizoctonia* populations of two field soils in the Adelaide Plains region of South Australia.

2. 原句：A balance between deep and shallow rooting plants, heavy and light feeders, nitrogen fixers and consumers and an undisturbed phase is needed to achieve maximum benefit through rotation.

改写后：Maximum benefit through rotation can be achieved by using a balance between deep and shallow rooting plants, heavy and light feeders, nitrogen fixers and consumers and an undisturbed phase. 还可以改写成：

To achieve maximum benefit through rotation, it is necessary to have a balance between deep and shallow rooting plants, heavy and light feeders, nitrogen fixers and consumers and an undisturbed phase.

任务 9.2 讨论部分的内容要点

见表 AP14、表 AP15 和表 AP16。

表 AP14 任务 9.2: PEA 1 讨论部分的内容要点

讨论部分	内容要点
The competition experiments shown in Figure 5(c) indicate that GmDmt1 can transport other divalent cations in addition to ferrous iron. Zinc, copper and manganese all inhibited iron uptake. The ability of GmDmt1;1 to enhance growth of the *zrt1zrt2* yeast mutant further suggests that the protein is not specific for iron transport.	2a. Restatement of one of the main findings, showing how it contributes to the main activity of the study
The preferred substrate *in vivo* may well depend on the relevant concentrations of divalent metals in the infected cell cytosol.	3. Speculation about the finding
This lack of specificity has been found with Nramp homologues from other organisms, including Nramp2 from mice. Despite this lack of specificity when expressed in heterologous systems, mutation of murine Nramp2 results in an anaemic phenotype, demonstrating that *in vivo* it is predominantly an iron transporter (Fleming et al. 1997).	2b. Comparison with the findings of other researchers
Although GmDmt1;1 was able to complement the DEY1453(*fet3fet4*) yeast mutant, the complementation was not robust and the growth media had to	2a. Continued review of the finding

续表

讨论部分	内容要点
be supplemented with low concentrations of iron. AtIrt1, on the other hand, showed much better complementation and allowed growth of the mutant in the absence of added iron(Figure 4). There are several possible reasons for the poorer growth with GmDmt1; 1, including possible instability of GmDmt1; 1 transcripts(perhaps because of the presence of the regulatory IRE element in the transcript).	3. Speculation about the findings

表 AP15 任务 9.2: PEA 2 讨论部分的内容要点

讨论部分	内容要点
Our experimental results demonstrate that space-and propagule-limitation both regulate *S. muticum* recruitment.	2a. Restatement of the most important finding showing how it contributes to the main activity of the study
Our finding that *S. muticum* recruitment was positively related to propagule input is similar to those of two previous studies(Parker 2001; Thomsen et al. 2006), in which the propagule input of invasive plants was manipulated. In our control treatment space was limiting, a result that has also been found in previous studies of *S. muticum* recruitment(Deysher & Norton 1982; De Wreede 1983; Sanchez & Fernandez 2006).	2b. Comparisons with the findings of other researchers
Consequently, increasing propagule pressure had a relatively weak effect on recruitment in undisturbed plots(Fig. 1a). However, when space limitation was alleviated by disturbing the plots, increasing propagule pressure caused a dramatic increase in recruitment(Fig. 1a).	2a. Continued review of the important findings
This suggests that in the presence of adequate substratum for settlement, propagule limitation becomes the primary factor controlling *S. muticum* recruitment. These results indicate that *S. muticum* recruitment under natural field conditions will be determined by the interaction between disturbance and propagule input.	5. Implications of the results(what they mean in the context of the broader field)

表 AP16 任务 9.2: PEA 3 讨论部分的内容要点

讨论部分	内容要点
The uncertainty in satellite derived TADR estimates is quite large, up to about 50%, but it is comparable to the error in field-based effusion rate measurements(Calvari et al. 2003; Harris & Baloga 2009; Harris & Neri 2002; Harris et al. 2007; Sutton et al. 2003).	2a and 2b. Restatement of a finding and comparison with the findings of other researchers
The main uncertainties arise from the lack of constraint on the lava parameters used to convert thermal flux to TADR(Harris & Baloga 2009).	3. Explanation of findings, supported by citation
Moreover, the presence of ash strongly affected the satellite-based estimates of lava discharge rate. This led to an underestimation of the satellite-derived final volume, and to a difference in the timing of simulated lava flow emplacement.	3. Explanation of findings(based on the study conditions)
Moreover, a certain discrepancy between actual and modeled flow areas is to be expected, as changes in the contemporaneous effusion rate take time to be translated into a change in active flow area(that is, the active flow area at any one time is a function of the antecedent effusion rate, rather than the instantaneous effusion rate)(Harris & Baloga 2009; Wright et al. 2001).	3. Explanation of findings, supported by citation

任务 9.5 利用动词调整观点的强烈程度

见表 AP17。

表 AP17 任务 9.5: 利用动词调整观点的强烈程度

The presence of an IRE motif	implies suggests provides evidence indicates shows demonstrates	that GmDmt1;1 mRNA	might be stabilized could be stabilized may be stabilized can be stabilized was stabilized should be stabilized is stabilized	by the binding of IRPs in soybean nodules when free iron levels are low.	Weak ↓ Strong

说明:

- 观点的强烈程度是非常主观的，在排序问题上英语为母语的作者也难以达成统一意见。表 AP17 中给出的参考答案是基于多年来我们举办研讨会的讨论结果，而设计这道习题的主要目的就是希望能引起你对此类用法的重视。在阅读自己领域的论文时，一定要关注相关的动词和形式，体会观点的强度。
- “can” 表示本次实验出现了当前结果，并记录在案，说明了当前结果出现的可能性。但无法确定该结果在未来是否会重复出现。
- “was stabilized” 不对当前结果在其他实验或研究条件下进行任何推广（如果从句前半段使用 “was stabilized”，那么后半部分也需要改成过去时态，即 “when free iron levels were low” ）。
- “should” 表示当前结果重复出现的可能性极高，尤其是具备某种条件的情况下（例如本句的 “when free iron levels are low” ）。“should” 还有一种用法表示建议他人做某事（例如：“You should wash your hands before meals” .），这在科技文体中并不常见，但是我们在 PEA 2 的讨论部分能找出一例：“The model results demonstrate that caution *should be exercised* when extrapolating the results of short-term disturbance experiments over longer time intervals” . 但是注意，使用这种用法提出建议需要预先充分说明理由。具体到 PEA 2 中的例子，前文已经详细讨论过模拟结果，在有论据支撑的前提下，作者才使用 should 提出建议。

任务 10.1 分析论文的标题

见表 AP18。

表 AP18 任务 10.1: 分析论文的标题

问题	PEA1	PEA2	PEA3
Is the title a noun phrase, a sentence, or a question?	Sentence	Noun phrase	Noun phrase

续表

问题	PEA1	PEA2	PEA3
How many words are used in the title?	16	13	15
What is the first idea in the title?	"The soybean NRAMP homologue, GmDMT1": the descriptor and name of the transporter discovered	"Short- and long-term effects"	"An emergent strategy"
Why do you think this idea has been placed first?	The descriptor comes first to show how this new discovery relates to what was previously known about the system under study.	This phrase highlights what is new and important about the work being reported.	This phrase indicates the status of the work being reported (not finalized).

任务 11.1 分析论文的摘要

见表 AP19、表 AP20 和表 AP21。

表 AP19 任务 11.1: 分析论文 PEA1 的摘要

论文摘要	内容要点
Iron is an important nutrient in N_2-fixing legume root nodules. Iron supplied to the nodule is used by the plant for the synthesis of leghemoglobin, while in the bacteroid fraction, it is used as an essential cofactor for the bacterial N_2-fixing enzyme, nitrogenase, and iron-containing proteins of the electron transport chain. The supply of iron to the bacteroids requires initial transport across the plant-derived peribacteroid membrane, which physically separated bacteroids from the infected plant cell cytosol.	Background
In this study, we have identified *Glycine max divalent metal transporter* (*GmDmt*1), a soybean homologue of the NRAMP/Dmt1 family of divalent metal ion transporters.	Principal activity
*GmDmt*1 shows enhanced expression in soybean root nodules and is most highly expressed at the onset of nitrogen fixation in developing nodules.	Results①
Antibodies raised against a partial fragment of GmDmt1	Method
confirmed its presence on the peribacteroid membrane (PBM) of soybean root nodules. GmDmt1 was able to both rescue growth and enhance $^{55}Fe(II)$ uptake in the ferrous iron transport deficient yeast strain (*fet*3*fet*4).	Results①
The results indicate that GmDmt1 is a nodule-enhanced transporter capable of ferrous iron transport across the PBM of soybean root nodules.	Conclusion
Its role in nodule iron homeostasis to support bacterial nitrogen fixation is discussed.	Another activity of the study/paper

① 写到研究结果的第一句话使用了一般现在时，言外之意：使用该方法得出的这些结果“永远为真”；研究结果的第三句话使用了一般过去时，说明这些研究结果是在特定实验条件下得出的（本次的实验结果如此）。

表 AP20 任务 11.1: 分析论文 PEA2 的摘要

论文摘要	内容要点
1. Invading species typically need to overcome multiple limiting factors simultaneously in order to become established, and understanding how such factors interact to regulate the invasion process remains a major challenge in ecology.	Background
2. We used the invasion of marine algal communities by the seaweed *Sargassum muticum* as a study system to experimentally investigate the inde-	Method + principal activity 1

论文摘要	内容要点
pendent and interactive effects of disturbance and propagule pressure in the short term. Based on our experimental results, we parameterized an integrodifference equation model, which we used to examine how disturbances created by different benthic herbivores influence the longer term invasion success of *S. muticum*.	Method + principal activity 2
3. Our experimental results demonstrate that in this system neither disturbance nor propagule input alone was sufficient to maximize invasion success. Rather, the interaction between these processes was critical for understanding how the *S. muticum* invasion is regulated in the short term.	Results
4. The model showed that both the size and spatial arrangement of herbivore disturbances had a major impact on how disturbance facilitated the invasion, by jointly determining how much space-limitation was alleviated and how readily disturbed areas could be reached by dispersing propagules.	Results
5. Synthesis. Both the short-term experiment and the long-term model show that *S. muticum* invasion success is co-regulated by disturbance and propagule pressure. Our results underscore the importance of considering interactive effects when making predictions about invasion success.	Results summary Conclusion/ recommendation

表 AP21 任务 11.1: 分析论文 PEA3 的摘要

论文摘要	内容要点
Spaceborne remote sensing techniques and numerical simulations have been combined in a web-GIS framework (LAV @ HAZARD) to evaluate lava flow hazard in real time.	Method + Purpose/ principle activity
By using the HOTSAT satellite thermal monitorng system to estimate time-varying TADR (time averaged discharge rate) and the MAGFLOW physics-based model to simulate lava flow paths,	Method
the LAV@HAZARD platform allows timely definition of parameters and maps essential for hazard assessment, including the propagation time of lava flows and the maximum run-out distance.	Results/Conclusion
We used LAV @ HAZARD during the 2008-2009 lava flow-forming eruption at Mt Etna (Sicily, Italy). We measured the temporal variation in thermal emission (up to four times per hour) during the entire duration of the eruption using SEVIRI and MODIS data.	Method
The time-series of radiative power allowed us to identify six diverse thermal phases each related to different dynamic volcanic processes and associated with different TADRs and lava flow emplacement conditions.	Result
Satellite-derived estimates of lava discharge rates were computed and integrated for the whole period of the eruption (almost 14 months),	Method
showing that a lava volume of between 32 and 61 million cubic meters was erupted of which about 2/3 was emplaced during the first 4 months.	Result
These time-varying discharge rates were then used to drive MAGFLOW simulations to chart the spread of lava as a function of time.	Method
TADRs were sufficiently low (b30 m^3/s) that no lava flows were capable of flowing any great distance so that they did not pose a hazard to vulnerable (agricultural and urban) areas on the flanks of Etna.	Conclusion

任务 12.1 分析综述的摘要

见表 AP22 和表 AP23。

表 AP22 任务 12.1: 分析 Harris et al. (2011) 综述的摘要

综述摘要	对应的内容要点
Plant development is adapted to changing environmental conditions for optimizing growth. This developmental adaptation is influenced by signals from the environment, which act as stimuli and may include submergence and fluctuations in water status, light conditions, nutrient status, temperature and the concentrations of toxic compounds. The homeodomain-leucine zipper(HD-Zip) I and HD-Zip II transcription factor networks regulate these plant growth adaptation responses through integration of developmental and environmental cues.	Background (information generally available and accepted)
Evidence is emerging that these transcription factors are integrated with phytohormone-regulated developmental networks, enabling environmental stimuli to influence the genetically preprogrammed developmental progression. Dependent on the prevailing conditions, adaptation of mature and nascent organs is controlled by HD-Zip I and HD-Zip II transcription factors through suppression or promotion of cell multiplication, differentiation and expansion to regulate targeted growth. In vitro assays have shown that, within family I or family II, homo- and/or heterodimerization between leucine zipper domains is a prerequisite for DNA binding. Further, both families bind similar 9-bp pseudopalindromic cis elements, CAATNATTG, under in vitro conditions.	Results(information representing new synthesis)
However, the mechanisms that regulate the transcriptional activity of HD-Zip I and HD-Zip II transcription factors in vivo are largely unknown. The in planta implications of these protein-protein associations and the similarities in cis element binding are not clear.	Conclusion/recommendation(highlighting remaining gaps in knowledge)

表 AP23 任务 12.1: 分析 Wymore et al. (2011) 综述的摘要

综述摘要	对应的内容要点
Genes and their expression levels in individual species can structure whole communities and affect ecosystem processes.	Background(information generally available and accepted)
Although much has been written about community and ecosystem phenotypes with a few model systems, such as poplar and goldenrod, here we explore the potential application of a community genetics approach with systems involving invasive species, climate change and pollution.	Purpose and scope(including the point of differentiation from previous work)
We argue that community genetics can reveal patterns and processes that otherwise might remain undetected. To further facilitate the community genetics or genes-to-ecosystem concept, we propose four community genetics postulates that allow for the conclusion of a causal relationship between the gene and its effect on the ecosystem. Although most current studies do not satisfy these criteria completely,	Results (information representing new synthesis/conclusions)
several come close and, in so doing, begin to provide a genetic-based understanding of communities and ecosystems, as well as a sound basis for conservation and management practices.	Conclusion/recommendation(claim for significance of the new synthesis)

任务 12.2 分析综述引言的结尾

见表 AP24 和表 AP25。

表 AP24 任务 12. 1: 分析 Harris et al. (2011) 综述引言的结尾

引言的结尾出现了层级 4 的哪些内容	标志词
b	We present ...; we ... describe ... and indicate ... and address the issue of ...
是否出现了层级 6: map of subsequent arrangement	标志词
是	• current knowledge that relates to the roles of the HD-Zip I and HD-Zip II TFs during plant adaptation under changing environmental conditions • how these TFs integrate with phytohormone-mediated responses • the limits of the current knowledge that relate to the mechanism of transcriptional activity • how to overcome these limitations

表 AP25 任务 12. 1: 分析 Wymore et al. (2011) 综述引言的结尾

引言的结尾出现了层级 4 的哪些内容	标志词
a	The major goal of this review is to explore...
b	We develop our ideas ...; we explore ...; we explore ...; we explore ...
c	Thus, ... can broaden our understanding of ... and remind us of ...
是否出现了层级 6: map of subsequent arrangement	标志词
是	• ... in conifers, ... how the interactions of foundation species (trees and squirrels) and climate can affect a much larger community • how a single mutation in one example and a single haplotype in another example can have cascading effects to redefine their respective ecosystems • how pollution can alter the gene expression of foundation species, which, in turn, may redefine these ecosystems

任务 13. 1 利用投稿信推荐自己的论文

见图 AP1 中突出显示的内容。

Please find attached the manuscript “Arbuscular mycorrhizal associations of the southern Simpson Desert”. This manuscript examines the mycorrhizal status of plants growing on the different soils of the dune-swale systems of the Simpson Desert. **There have been few studies** of the ecology of the plants **in this desert** and **little is known about** how mycorrhizal associations are distributed amongst the desert plants of Australia. We report the arbuscular mycorrhizal status **of 47 plant species for the first time.** The manuscript has been prepared according to the journal’s Instructions for Authors. We believe that this **new** work is within the scope of your journal and hope that you will consider this manuscript for publication in the *Australian Journal of Botany*.

图 AP1 任务 13. 1: 论文作者使用了哪些词组向编辑“强烈推销”自己的研究

任务 17.1 常见的错误类型

问题 2：常见错误类型及其严重程度（见表 AP26）

可供选择的错误类型：

1. 单复数形式有误（例如：all tea leaves sample were oven dried）；

2. 句式过于复杂或不够准确（例如：This may be due to lower pH hinders dissolution of soil organic matter and decreases total dissolved Cu concentration because of Cu-organic complex reducing.）；

3. 主谓不一致（例如：the results of this study suggests that …）；

4. 介词使用有误（例如：similar with the results of other researchers）；

5. 冠词 a/an/the 使用不当（例如：the accumulation of Cu in human body）；

6. 情态动词使用有误（例如：would 和 will 的误用，can，could 和 may 的误用）；

7. 词性有误（例如：drought resistance varieties）；

8. 时态选用不合理（例如：描述已经得出的实验结果时使用现在时态）。

表 AP26 任务 17.1：常见错误类型及其严重程度

Rarely/slightly affects meaning	Sometimes/moderately affects meaning	Often/seriously affects meaning
1 3	4 5 7	2 6 8

说明：表 AP26 给出的答案不是绝对的。我们分析了大量母语非英语作者写出的文章，这些审稿经验是我们做出上述分类的依据。除此之外，还有两点值得注意：

- 上面所有的错误类型都会对语义构成一定程度的影响，需要结合实例具体问题具体分析；
- 即使是那些基本不影响语义的错误也可能会激怒读者，给人留下负面印象，读者可能会基于这些不准确的语言而质疑论文内容（科学层面）的准确性。

问题 3：如何改正这些错误

错误（1）、（4）、（7）可以借助 AdTAT（详见 17.5 节）等语料库检索软件来进行纠正，AdTAT 对改正错误（2）和（6）有时也有帮助。错误（5）冠词问题可以借助 17.6 节介绍的图 17.1 来解决。在文中修正错误（1）和（3）时，可以使用 15.2 节第 7 条介绍的方法，借助直尺和论文的纸质版逐行复查。

任务 17.2 为引言的层级 4 制作句子模板

PEA2

In this study we used [NP1] as a study system to better understand the effects of [NP2] and [NP3] on [NP4]. In a [adjectives] experiment we manipulated both [NP2] and [NP3] in order to examine how these factors [adverbs] influence [NP5] in [NP6]. We supplement the experimental results with [NP7], which we use to examine how [NP8] influence [NP9] in [NP10].

PEA1

In this study we have identified [NP1], [NP1a]. We show that [NP1a] is [NP2], expressed in [NP3] at [NP4], and is localised to [NP5]. [NP1a] is capable of [NP6] when expressed in [NP7].

PEA3

Here we present [NP1], named [NP2], which integrates [NP3] with [NP4] to simulate [NP5]. ... Here we describe and demonstrate the operation of [NP2] using a [adjective] analysis of [NP6].

任务 17.6 表示泛指的名词短语

Legumes form symbiotic associations with N_2-fixing soil-borne bacteria of the *Rhizobium* family. The symbiosis begins when compatible bacteria invade legume root hairs, signalling the division of inner cortical root cells and the formation of a nodule. Invading bacteria migrate to the developing nodule by way of an 'infection thread', comprised of an invaginated cell wall. In the inner cortex, bacteria are released into the cell cytosol, enveloped in a modified plasma membrane (the peribacteroid membrane (PBM)), to form an organelle-like structure called the symbiosome, which consists of bacteroid[①], PBM[①] and the intervening peribacteroid space (PBS; Whitehead and Day, 1997). The bacteria, subsequently, differentiate into the N_2-fixing bacteroid form. The symbiosis allows the access of legumes to atmospheric N_2, which is reduced to NH_4^+ by the bacteroid enzyme nitrogenase. In exchange for reduced N, the plant provides carbon to the nodules to support bacterial respiration, a low-oxygen environment in the nodule suitable for bacteroid nitrogenase activity, and all the essential nutritional elements necessary for bacteroid activity. Consequently, nutrient transport across the PBM is an important control mechanism in the promotion and regulation of the symbiosis.

① 这两个词可以看作可数名词，但是作者在这里将它们用作不可数名词，表示不同的组织（tissue types）。

冠词确实非常复杂，专家们甚至也会对它们在具体语境下的用法争论不休。想要了解自己领域内冠词是如何使用的，最好的方法就是使用 AdTAT 进行学习。

任务 17.7 表示特指的名词短语

突出显示（灰色底色）的内容是表示特指的名词短语。

Legumes form symbiotic associations with N_2-fixing soil-borne bacteria of the *Rhizobium* family. The symbiosis begins when compatible bacteria invade legume root hairs, signalling the division of inner cortical root cells and the formation of a nodule. Invading bacteria migrate to the developing nodule by way of an 'infection thread', comprised of an invaginated cell wall. In the inner cortex, bacteria are released into the cell cytosol, enveloped in a modified plasma membrane (the peribacteroid membrane (PBM)), to form an organelle-like structure called the symbiosome, which consists of bacteroid, PBM and the intervening peribacteroid space (PBS; Whitehead and Day, 1997). The bacteria, subsequently, differentiate into the N_2-fixing bacteroid form. The symbiosis allows the access of legumes to atmospheric N_2, which is reduced to NH_4^+ by the bacteroid enzyme nitrogenase. In exchange for reduced N, the plant provides carbon to the nodules to support bacterial respiration, a low-oxygen environment in the nodule suitable for bacteroid nitrogenase activity, and all the essential nutritional elements necessary for bacteroid activity. Consequently, nutrient transport across the PBM is an important control mechanism in the promotion and regulation of the symbiosis.

任务 17.8 科技文体中的冠词和名词复数

Propagule pressure is widely recognized as **an** important factor that influences invasion success. Previous studies suggest that **the** probability of successful invasion increases with **the** number of propagules released, with **the** number of introduction attempts, with introduction rate, and with proximity to existing populations of invaders. Moreover, propagule pressure may influence invasion dynamics after establishment by affecting **the** capacity of non-native species to adapt to their new environment. Despite its acknowledged importance, propagule pressure has rarely been manipulated experimentally and **the** interaction of propagule pressure with other processes that regulate invasion success is not well understood.

注意：文中的“propagule pressure”是泛指概念，且“pressure”在这里也不可数，因此不需要添加冠词，类似的情况还有“introduction rate，proximity”和“invasion success”。

任务 17.9 添加标点符号

1. Lime，which raises the pH of the soil to a level more suitable for crops，is injected into the soil using a pneumatic injector.

2. 不需要添加标点。

3. Non-cereal phases，which are essential for the improvement of soil fertility，break disease cycles and replace important soil nutrients.

4. Senescence，which is the aging of plant parts，is caused by ethylene that the plant produces.

5. 不需要添加标点。

6. Seasonal cracking，which is a notable feature of this soil type，provides pathways at least 6mm wide and 30cm deep that assist in water movement into the subsoil.

7. 不需要添加标点。

8. Yellow lupin，which may tolerate waterlogging better than the narrow-leafed variety，has the potential to improve yields in this area.

9. 不需要添加标点。

附录 期刊的质量和影响力

说到一段时间内期刊的质量或其学术价值，其实没有什么简便的衡量方法。人们提出了许多引证指标，试图评价期刊文章的引用量以及在学术界引起反应的速度。当然，这些引证指标能够从一定程度上体现期刊的总体水平，其中最常用的指标当属影响因子（Journal Impact Factor）。

附录 1 期刊的影响因子

期刊某一特定年份的影响因子是该年引证该刊前 2 年论文的总次数与前 2 年该刊所发表的论文总数之比，度量的是该刊近期发表论文的平均被引频次。计算公式如下：

$$x\text{ 年的影响因子}=\frac{\text{在}(x-1)\text{年与}(x-2)\text{年文章被引总次数}}{\text{在}(x-1)\text{年与}(x-2)\text{年发表文章总数}}$$

评价期刊影响力的其他引证指标还包括以下四项。

- 五年影响因子（5-year Journal Impact Factor）：计算影响因子时采用的是 5 年数据，对上述公式稍作调整即可。
- 即年指标（Journal Immediacy Index）：某期刊当年发表论文的被引次数除以该刊当年发表的论文数。指标值越高，说明该刊论文在学术界引起反响的速度越快。
- 被引半衰期（Journal Cited Half-Life）：某期刊论文在某年被引用的总次数中，较新的一半是距离现在多长一段时间内发表的，这是衡量期刊影响力久远程度的指标。
- 特征因子分值（Eigenfactor® Score）：计算时考虑的是 5 年引用数据，而且同时考虑了施引期刊的影响力，高影响力的施引期刊在计算中所占的权重也越高。

附录 2 正确看待期刊质量的评价指标

使用不同的引证指标时，需要了解每种指标的设计目的和其局限性。我们在上面介绍过的指标都是针对期刊中平均每篇文章的引用速率和数量而言的，因而它们评价的是期刊的整体影响力，而非个别文章的质量。你可以用类似的方法去计算自己发表的论文，得出的结果可能会高于或低于期刊的平均值。论文唯有出现在正确的读者面前，才有可能被阅读或被引用。有些时候，目标读者关注的未必是影响因子最高的期刊。

因此，在利用各种评价指标对期刊进行排序时，还需要考虑下列因素：

- 不同学科期刊的评价指标缺乏可比性（例如：分子生物学领域的论文通常要引用大量文献，而数学学科则不是这样）。

- 如果期刊除了发表原创性研究，还包含综述或其他类似专栏，那么一些评价指标在计算时实际上夸大了期刊的相对价值。
- 与原始文献进行匹配的过程也受各种因素影响，例如参考文献的信息或标引不准确、期刊的校对与纠错流程、期刊采用的引用规范、作者姓名匹配与识别、语言问题以及不熟悉某些国家使用的人名等。
- 论文被引次数的统计往往局限于特定的期刊目录，这个目录很大程度上排除了在母语非英语的国家出版的期刊，以及一些没有名气的新期刊。
- 根据评价指标得出的期刊排名是不断变化的。图 A1 展现的就是植物学领域内三种常见期刊影响因子的变化。可以看出，十年间有的期刊的影响因子相对稳定，有的逐渐攀升，有的则先降后升，但是三种期刊中那些优质的论文依然会继续被人引用。

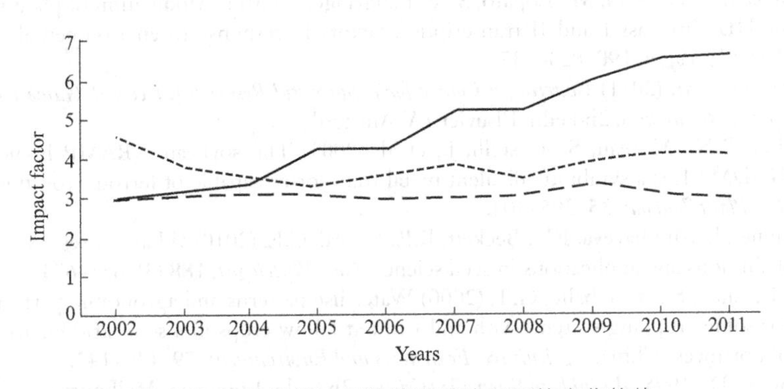

图 A1 植物学领域三种期刊影响因子的变化趋势

- 具体到你发表的论文，其学术质量是由多种因素决定的，而不仅仅是片面追求期刊的影响因子。随着开放存取等方式的兴起，读者已经可以从多种渠道获取高质量的科技文献，这就需要你正确看待期刊质量的评价指标。期待你能在科研生涯中发表更多的优质研究，为自己的研究领域不断贡献有价值的内容。

参考文献

[1] Britton-Simmons, K.H. & Abbott, K.C. (2008) Short- and long-term effects of disturbance and propagule pressure on a biological invasion. *Journal of Ecology*, **96**, 68–77.

[2] Burgess, S. & Cargill, M. (2013) Using genre analysis and corpus linguistics to teach research article writing. In: *Supporting Research Writing: Roles and Challenges in Multilingual Settings* (ed. V. Matarese), pp. 55–71. Woodhead Publishing, Cambridge.

[3] Burrough-Boenisch, J. (1999) International reading strategies for IMRD articles. *Written Communication*, **16** (3), 296–317.

[4] Cadman, K. & Cargill, M. (2007) Providing quality advice on candidates' writing. In: *Supervising Doctorates Downunder: Keys to Effective Supervision in Australia and New Zealand* (eds C. Denholm & T. Evans), pp. 182–191. ACER, Melbourne.

[5] Flowerdew, J. & Li, Y. (2007) Language re-use among Chinese apprentice scientists writing for publication. *Applied Linguistics*, **28**, 440–465.

[6] Ganci, G., Vicari, A., Cappello, A., & Del Negro, C. (2012) An emergent strategy for volcano hazard assessment: from thermal satellite monitoring to lava flow modelling. *Remote Sensing of Environment*, **119**, 197–207.

[7] Harris, J.C., Hrmova, M., Lopato, S., & Langridge, P. (2011) Modulation of plant growth by HD-Zip class I and II transcription factors in response to environmental stimuli. *New Phytologist*, **190**, 823–837.

[8] Johnson, A.M. (2011) *Charting a Course for a Successful Research Career: A Guide for Early Career Researchers*, 2nd edn. Elsevier BV, Amsterdam.

[9] Kaiser, B.N., Moreau, S., Castelli, J., et al. (2003) The soybean NRAMP homologue, GmDMT1, is a symbiotic divalent metal transporter capable of ferrous iron transport. *The Plant Journal*, **35**, 295–304.

[10] Kranner, I., Minibayeva, F.V., Beckett, R.P., & Seal, C.E. (2010) What is stress? Concepts, definitions and applications in seed science. *New Phytologist*, **188** (3), 655–673.

[11] Li, F., Zhao, S., & Geballe, G.T. (2000) Water use patterns and agronomic performance for some cropping systems with and without fallow crops in a semi-arid environment of northwest China. *Agriculture, Ecosystems and Environment*, **79**, 129–142.

[12] Lindsay, D. (1995) *A Guide to Scientific Writing*, 2nd edn. Longman, Melbourne.

[13] Lindsay, D. (2011) *Scientific Writing = Thinking in Words*. CSIRO Publishing, Collingwood.

[14] McNeill, A.M., Zhu, C.Y., & Fillery, I.R.P. (1997) Use of *in situ* ^{15}N-labelling to estimate the total below-ground nitrogen of pasture legumes in intact soil-plant systems. *Australian Journal of Agricultural Research*, **48**, 295–304.

[15] Mullins, G. & Kiley, M. (2002) 'It's a PhD, not a Nobel Prize': how experienced examiners assess research theses. *Studies in Higher Education*, **27** (4), 369–386.

[16] Pechenik, J.A. (1993) *A Short Guide to Writing about Biology*, 4th edn. Addison Wesley Longman, New York.

[17] Sarpeleh, A., Wallwork, H., Catcheside, D.E.A., Tate, M.E., & Able, A.J. (2007) Proteinaceous metabolites from *Pyrenophora teres* contribute to symptom development of barley net blotch. *Phytopathology*, **97**, 907–915.

[18] Weissberg, R. & Buker, S. (1990) *Writing Up Research: Experimental Research Report Writing for Students of English*. Prentice Hall Regents, Englewood Cliffs.

[19] Wymore, A.S., Keeley, A.T.H., Yturralde, K.M., Schroer, M.L., Propper, C.R., & Whitham, T.G. (2011) Genes to ecosystems: exploring the frontiers of ecology with one of the smallest biological units. *New Phytologist*, **191** (1), 19–36.